Lecture Notes in Biomathematics

Managing Editor: S. Levin

76

Catherine A. Macken
Alan S. Perelson

Stem Cell Proliferation and Differentiation

A Multitype Branching Process Model

Springer-Verlag

Berlin Heidelberg New York London Paris Tokyo

Authors

Catherine A. Macken
Department of Mathematics and Statistics, University of Auckland
Private Bag, Auckland, New Zealand

Alan S. Perelson
Theoretical Division
Los Alamos National Laboratory
Los Alamos, NM 87545, USA

Mathematics Subject Classification (1980): 92A05

ISBN-13: 978-3-540-50183-1   e-ISBN-13: 978-3-642-93396-7
DOI: 10.1007/978-3-642-93396-7

Preface

The body contains many cellular systems that require the continuous production of new, fully functional, differentiated cells to replace cells lacking or having limited self-renewal capabilities that die or are damaged during the lifetime of an individual. Such systems include the epidermis, the epithelial lining of the gut, and the blood. For example, erythrocytes (red blood cells) lack nuclei and thus are incapable of self-replication. They have a life span in the circulation of about 120 days. Mature granulocytes, which also lack proliferative capacity, have a much shorter life span – typically 12 hours, though this may be reduced to only two or three hours in times of serious tissue infection. Perhaps a more familiar example is the outermost layer of the skin. This layer is composed of fully mature, dead epidermal cells that must be replaced by the descendants of stem cells lodged in lower layers of the epidermis (cf. Alberts et al., 1983). In total, to supply the normal steady-state demands of cells, an average human must produce approximately 3.7×10^{11} cells a day throughout life (Dexter and Spooncer, 1987). Common to each of these cellular systems is a primitive (undifferentiated) stem cell which replenishes cells through the production of offspring, some of which proliferate and gradually differentiate until mature, fully functional cells are produced. In most instances, the mature cells are incapable of further division, and, after a variable period of time, they die. The stem cell is defined by its capacity for infinite self-renewal. The series of events that lead from a stem cell offspring to a mature end cell is under the control of growth factors. The number of cells of a given type is maintained at equilibrium *in vivo* through a delicate balance between cell proliferation and cell differentiation, ultimately leading to cell death. Cell proliferation and differentiation within the blood is the subject of this monograph.

The hemopoietic (blood) stem cell has been difficult to isolate, although preliminary reports from the laboratory of I. Weissman (Stanford University) indicate that the stem cell may finally have been isolated by a procedure involving both positive and negative selection for cell surface markers (Muller-Sieburg et al., 1986). The existence of the stem cell was demonstrated in a series of famous experiments by Till and McCulloch (1961). They devised the spleen-colony assay which consisted of injecting a sample of bone marrow intravenously into mice whose hemopoietic systems had been destroyed by heavy irradiation. As evidenced by their continued survival, the injection of bone marrow cells enabled the mice to reconstitute their hemopoietic systems. After ten days the mice were killed and their spleens excised. (The spleen is a highly vascularized organ that provides a favorable environment for the growth and differentiation of bone marrow cells.) The spleens of mice injected with bone marrow cells were marked by macroscopic nodules that contained cells having the capacity to produce further spleen nodules, as well as histologically recognizable differentiated hemopoietic cells. Direct cytological evidence suggested that these nodules were cell colonies derived from single cells, i.e., were clones. Subsequent work established this more firmly. The single cells that gave rise to these nodules were given the operational name of "colony-forming cells" or "colony-forming units-spleen" (CFU-S). The conclusion that colony-forming cells were related to stem cells was made on the basis of three criteria: (1) the cells possessed extensive proliferative capacity since they were able within 10 days to generate colonies containing in excess of 10^6 cells, (2) they were capable of differentiation since the colonies contained large numbers of histologically recognizable differentiated cells, and (3) the cells were capable of self-renewal since the colonies contained cells that were themselves capable of forming spleen colonies. Further details of the Till and McCulloch experiment as well as an up-to-date introductory exposition on hemopoiesis can be found in Golub (1987).

Figure 1 shows conceptually the differentiation of a single hemopoietic stem cell into all the cellular components of the blood. However, the figure does not show all of the many stages of differentiation intermediate between the stem cell and a single fully differentiated end cell. Partially differentiated pluripotent cells have been identified that are capable of giving rise to various subsets of the complete set of component cells of the blood. For example, a single bipotential cell called the GM progenitor, or CFU-GM, has been shown to produce offspring that may be in either the granulocyte (G) or the macrophage (M) lineage (and hence the designation GM progenitor). Progenitors with a greater range of potentials have been isolated, such as the $GEMM$ progenitor, which can produce cells in the erythrocyte and megakaryocyte lineages, in addition to cells in the G and M lineages.

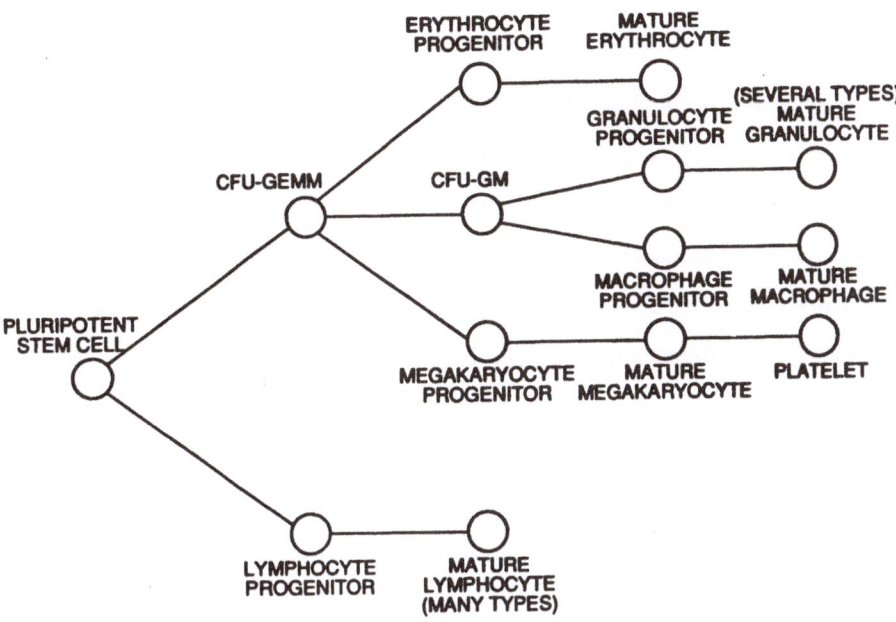

Figure 1. Schematic diagram of the pathways of hemopoietic cell differentiation. The pathways start with the hemopoietic pluripotent stem cell of the bone marrow. In this scheme differentiation is portrayed as occurring via a set of branch point decisions, with multipotent cells becoming committed to particular cell lineages.

In conjunction with identifying cells having greater or lesser capacity for differentiation, glycoproteins called growth factors, necessary for supporting the growth, differentiation and survival of these cells, have also been identified. Significantly, growth factors are required throughout development; if they are removed, the cells die (Dexter and Spooncer, 1987). Growth factors are generally labeled by the cell-lineages along which their target cells may differentiate. Thus, the growth factor supporting growth of the *GM* progenitor has been called GM-CSF. (CSF stands for colony stimulating factor.) An interleukin, IL-3, has been shown to support the growth of a primitive cell capable of producing colonies of mixed erythroid and myeloid forms including neutrophils, eosinophils, basophils, monocyte-macrophages, and megakaryocytes, and hence has been called multi-CSF (cf. Clark and Kamen, 1987; Groopman, 1987).

Much progress has been made on the identification of the various cell types and the multitude of growth factors that comprise the actors in the hemopoietic system. Understanding the dynamics of the interactions among the components of the system is very difficult. Clearly, fine-tuned feedback mechanisms must be involved in order to achieve the balance necessary for the maintenance of health. This study aims to introduce a general framework within which the processes of cell differentiation and proliferation may be studied. In order to achieve satisfactory levels of precision and realism in our modeling, we have chosen to focus on a single lineage of the differentiation process, namely, that lineage leading to macrophages. By studying this lineage in detail, we hope to lay the pathway to quantitative studies of wider aspects of hemopoietic stem cell proliferation and differentiation.

This work was performed under the auspices of the U.S. Department of Energy. It was supported in part by N.I.H. grant AI 19490. C.A.M. would like to thank the Theoretical Biology and Biophysics Group at Los Alamos National Laboratory for their hospitality during her many visits. We thank Simon Tavaré for helpful discussion on the problem of parameter estimation, and Jerry Nedelman and Jerome Percus for reading and commenting on an earlier version of the monograph. We are particularly grateful to Carleton C. Stewart for stimulating this research and for his many helpful comments and criticisms.

C. A. Macken

A. S. Perelson

March, 1988

Table of Contents

Preface

Chapter 1

Introduction

Our interest in modeling the process of cell growth and differentiation in cell culture systems was stimulated by *in vitro* experiments designed to study the growth and differentiation of bone marrow cells into mature macrophages and by attempts to maintain a line of macrophages in culture. When bone marrow or other not fully differentiated cells are grown on plates in the presence of nutrients and appropriate growth factors, some single cells grow into large groups of cells. Even when cells are grown under identical conditions, one finds great heterogeneity in the sizes of the resulting groups (see Fig. 1.1; also Fig. 1B of Sachs, 1987). As the length of time a plate is incubated increases, it is observed that the distribution of sizes of groups becomes bimodal (Stewart, 1980, 1984). Typically, groups are either relatively small ($\leq$ 50 cells) and are called *clusters*, or are orders of magnitude larger than clusters and are called *colonies*. As the culturing time increases further, fewer groups fall between these size limits. Pharr et al. (1985) observed a similar bimodality in the colony size distribution of proliferating mast cells.

Individual clusters or colonies can be dispersed and the constituent cells grown on new plates in new media. This process is called *subculturing*. Cells from clusters that are subcultured tend to produce new clusters or do not grow at all. In comparison, cells from colonies that are subcultured have a much greater chance of producing colonies that may themselves be subcultured. This contrasting behavior has suggested to some experimentalists that colonies grow from a parent with a high capacity for proliferation; whereas clusters more probably grow from a more highly differentiated parent with limited growth potential. Our model, as developed in this monograph, is in part aimed at examining this conjecture. If there is a correlation between colony size and the likelihood of a colony containing a stem cell, then subculturing colonies may be an important facet of identifying and characterizing stem cells.

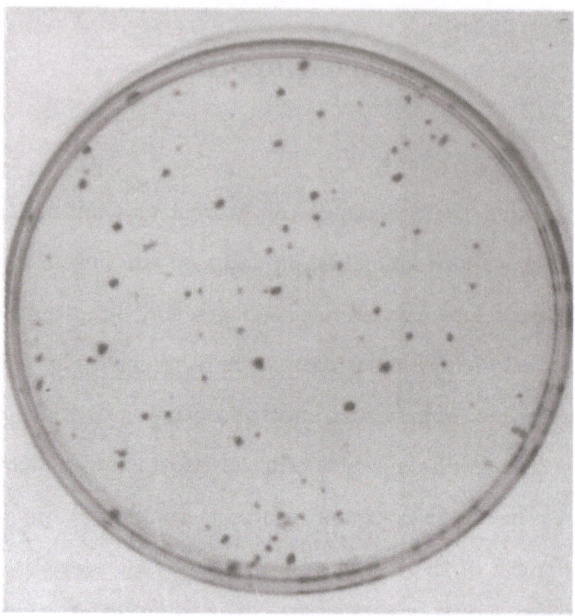

Figure 1.1   Colonies formed from murine mononuclear phagocyte progenitor cells grown in medium containing serum and macrophage growth factor. See Stewart (1984) for further experimental details. Notice the great heterogeneity in size among the various colonies.

When a colony or cluster will not grow further, we say that it has *reached completion*. In our model, completion results from the terminal differentiation of the proliferating cells in a colony into "end cells." The hypothesis that the cessation of proliferation in culture represents a step of differentiation is not new (cf. Bell et al., 1978) but is developed here within the context of a complete description of colony growth.

Our model of cellular differentiation and proliferation is based on the stochastic theory of multitype branching processes. The use of a stochastic model reflects our belief that the mechanism governing the decision of a cell to differentiate is essentially random, although it can be influenced by environmental agents such as growth factors and hormones. Further, from time lapse studies of cells grown in culture, family trees can be constructed (see Fig. 1.2)

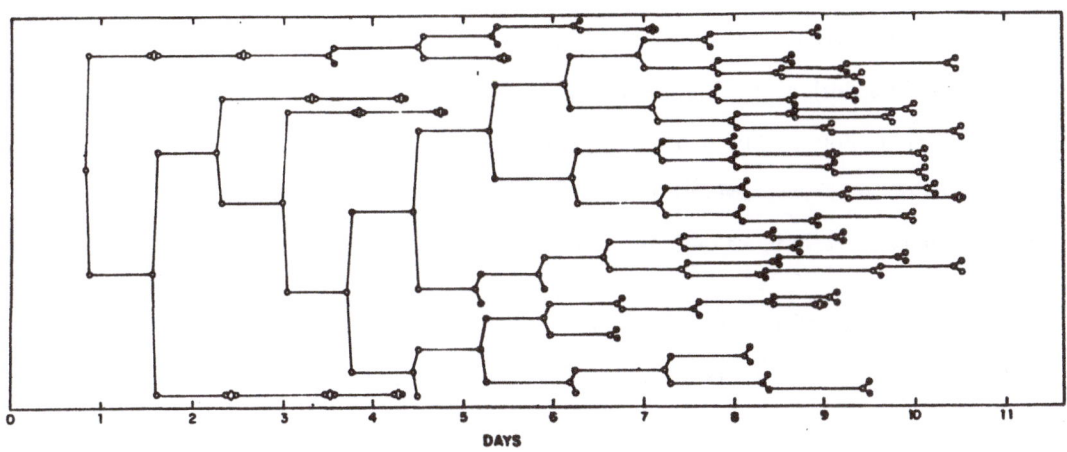

Figure 1.2   The growth of a single cell in culture was followed by time lapse photography. Closed circles represent cells that stopped proliferating. Open circles represent cells that continued to proliferate. The fate of cells after the last generation illustrated is not known. Cells that divide but fail to undergo cytokinesis are shown as giving rise to a single daughter cell. See Stewart (1980) for details.

that resemble the family trees of a classical Galton-Watson process (cf. Absher and Absher, 1976; Stewart, 1980; Matsumura, 1983, 1984). Motivated by this resemblance, we analyze the variability of cell growth seen in culture by constructing a branching process model of cell proliferation and differentiation. Our model takes into account three classes of cells: *stem cells*, with an infinite capacity for self-renewal; *progenitor cells*, with a finite capacity for self-renewal; and *end cells*, which are mature fully differentiated cells that do not divide. The branching process model predicts a high degree of variability in the cellular composition of growing colonies and clusters, in the size of colonies and clusters upon completion of growth, and in the time to completion of growth. Variability in all of these features of colony growth is observed experimentally. Using our theory we also predict the size of a colony as a function of time and develop criteria for determining if a colony will cease growing. For those colonies that stop growing, we compute the distribution of generation times until cessation of growth.

We also compute the distribution of ultimate colony sizes. Under some circumstances a colony can grow without bound. For such colonies we compute the asymptotic proportions of cell types within the colony. Our approach allows us to predict the outcome of rare but highly significant events, such as the presence of a stem cell in a colony, and allows us to summarize in a quantitative fashion the outcomes of large numbers of experimental observations. Practical applications of our theory are also obtained: we describe procedures for enriching for stem cells and replating techniques designed to ensure continued growth of cells in culture.

Both experimentalists and theoreticians have recognized that stochastic models can provide a means of describing the behavior of individual cells and their progeny. In order to provide some frame of reference for our modeling effort, we outline below the works of other authors selected for their relevance to the approach we adopt. Many other theoretical studies may be of interest to the reader, such as those of Kirkwood and Holliday, 1975; Good, 1977; Bjerknes, 1985, 1986; Wichmann and Loeffler, 1985; Cowan and Morris, 1986; and Day, 1986. (These references include studies of cell proliferation and differentiation systems other than the hemopoietic system.)

The earliest stochastic model of cell proliferation and differentiation known to us was developed by Till et al. (1964) in order to analyze data obtained with their classic spleen colony assay (Till and McCulloch, 1961; see also Preface). Till et al. noticed that the number of colony-forming cells varied greatly among colonies with just a few colonies containing a large number of colony-forming (i.e., stem) cells. Examining the frequency distribution of colony-forming cells in detail, they found that the variance in the number of stem cells greatly exceeded the mean, suggesting that the variation from colony to colony is greater than would be expected on the basis of sampling errors alone. Examining various empirical distributions, they found that the distribution of stem cells per colony did not differ significantly from a gamma distribution. In order to understand how such a distribution could arise, they modeled the proliferation of a stem cell as a continuous-time stochastic birth and death process (see Fig. 1.3a). A "birth" corresponded to the production of two colony-forming cells by division

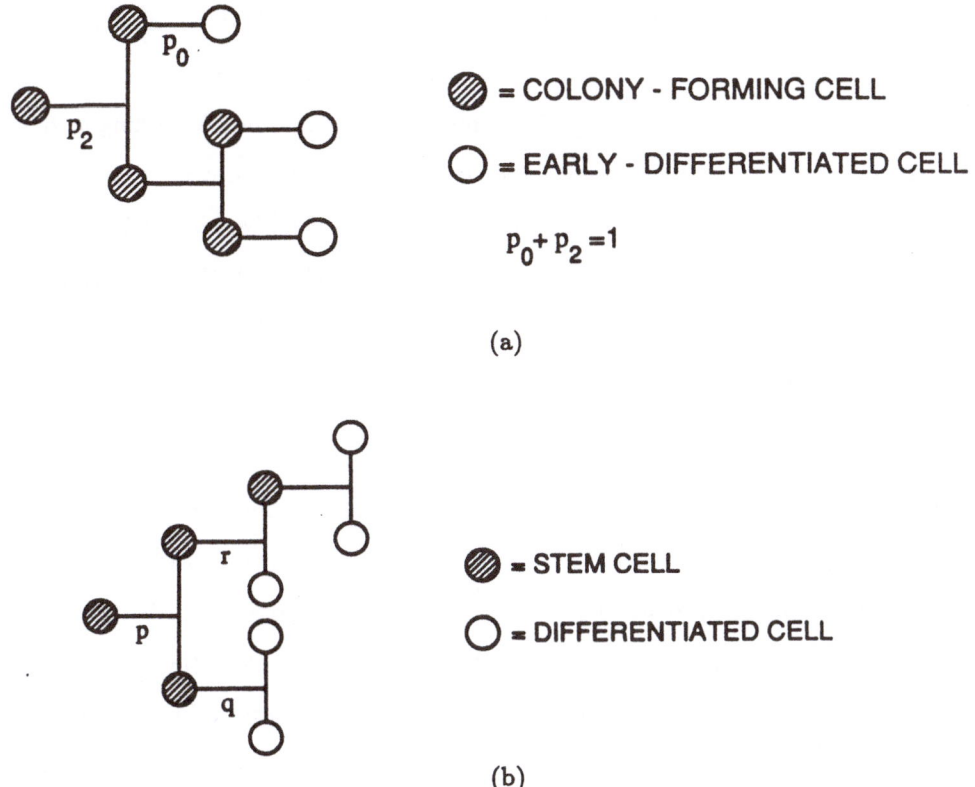

P_0

P_2

⬣ = COLONY - FORMING CELL

◯ = EARLY - DIFFERENTIATED CELL

$P_0 + P_2 = 1$

(a)

r

p

q

⬣ = STEM CELL

◯ = DIFFERENTIATED CELL

(b)

Figure 1.3   Examples of colony growth under different stochastic branching models. Cell types in each case are represented by: ⬣, earliest progenitor or stem cell; ◯, differentiated offspring. Probabilities of all possible proliferation and/or differentiation outcomes are illustrated. Where no further branching is shown, the last cell type is the end cell of the model. (a) Model of Till et al. (1964). Single-type model of growth of colony-forming cells in the spleen. Production of a differentiated cell is regarded as "death." At each discrete time point, a colony-forming cell either divides into two colony-forming cells (p_2), or differentiates (p_0). (b) Model of Vogel et al. (1969). Single-type model of growth of stem cells in the spleen. At each discrete time point, a stem cell divides to give rise to two, one, or no stem cells, with probabilities p, q, r, respectively, where $p + q + r = 1$. The fate of differentiated cells is completely deterministic.

of a single colony-forming cell. "Death" corresponded to the loss of colony forming ability by a cell because of differentiation. An explicit solution to this birth-death model had been obtained previously by Harris (1959) for the case in which the cell generation times were randomly distributed and followed a negative exponential distribution. According to Harris' solution the number of stem cells should be distributed according to a negative binomial. This was of particular interest because the limiting form of a negative binomial distribution is a gamma distribution. The gamma distribution also fitted well to Monte Carlo simulations of the birth-death process, with fixed generation times, and the birth and death probabilities chosen as 0.6 and 0.4, respectively.

The model of Till et al. was generalized by Vogel et al. (1969) in the context of the development of erythroid colonies. In this model a stem cell was allowed to reproduce asymmetrically, contributing 0, 1, or 2 stem cells (and consequently 2, 1, or 0 erythroid progenitors) to the next generation with probabilities q, r, and p, respectively (see Fig. 1.3b). Under this so-called "p-q-r" model, Vogel et al. calculated the mean and variance of the number of stem cells as a function of generation number n. They also derived the probability of extinction of all the stem cells in a colony as a function of n and found its asymptotic value. Using data from spleen colony assays, Vogel et al. estimated the probability of stem cell self-renewal, i.e. the parameter p, as $p = 0.61(1 - r)$. For a model with symmetric divisions, $r = 0$ and $p = 0.61$, a value quoted in the later literature.

Another early model, which was also developed with reference to *in vivo* data from spleen colony assays, is that of Korn et al. (1973). These authors noted the great variability in the composition of spleen colonies when differentiated cells were classified as being either granulocytic or erythroid in nature. They developed a multitype branching process model (see Fig. 1.3c) in which at each division of a stem cell, both daughter cells were stem cells, erythroid precursors, or granulocytic precursors, with probabilities p_S, p_E and p_G, respectively ($p_S + p_E + p_G = 1$). Growth kinetics of an *average* colony were predicted under the assumption that the generation times were fixed for a given cell type. Using $p_S = 0.61$, a value taken from the work of Vogel et al. (1969), and other parameter values constrained

(c)

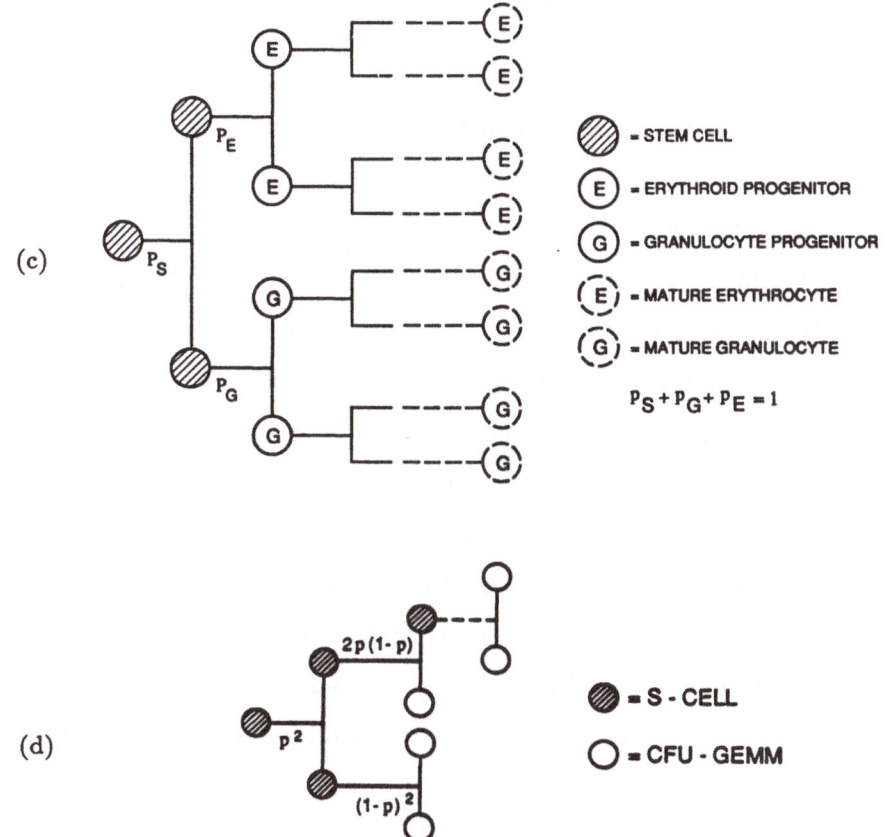

- STEM CELL

E - ERYTHROID PROGENITOR

G - GRANULOCYTE PROGENITOR

E - MATURE ERYTHROCYTE

G - MATURE GRANULOCYTE

$$P_S + P_G + P_E = 1$$

(d)

$2p(1-p)$

p^2

$(1-p)^2$

- S - CELL

- CFU - GEMM

Figure 1.3 continued. (c) Model of Korn et al. (1973). Multitype model of growth of stem cells in the spleen. At the end of each cell cycle, a stem cell divides into two cells of the same type, i.e. stem cells, erythroid progenitors, or granulocyte progenitors. Progenitors then divide and mature until a fully mature end state is reached. Cell cycle times may vary among cell types. (d) Model of Nakahata et al. (1982). Single-type model of growth in clonal assays. At each discrete time point, an S cell divides to give rise to two, one, or no S cells, with probabilities p^2, $2p(1-p)$, and $(1-p)^2$, respectively. Differentiated offspring are multipotent $CFU - GEMM$. Production of differentiated offspring is regarded as "death" for the purposes of the model.

by the data, they were able to fit their model to published experimental results on colony composition.

More recently, hemopoietic stem cell differentiation and proliferation has been studied *in vitro* by means of clonal assays of the type that we consider in our analysis of colonies formed in the macrophage lineage. Nakahata et al. (1982) identified in clonal cell cultures an undifferentiated blast cell that they believed to be more primitive than the so-called $CFU - GEMM$ (granulocyte-erythrocyte-macrophage-megakaryocyte colony-forming unit, i.e., a cell giving rise to colonies containing mixtures of these four cell types), the most primitive multipotential progenitor hitherto identified. To test their belief, Nakahata et al. formulated a branching process model similar to that of Till et al. (1964) (see Fig. 1.3d) in which the blast cell produced two daughter cells at the end of each cell cycle, 0, 1, or 2 of which may be blast cells, while the remaining daughter cells differentiated into a $CFU - GEMM$. Using fixed generation times, they showed that their model could adequately fit data on the observed frequency of blast cells and $CFU - GEMM$ in 16-day colonies, thus lending support to the proposal that the undifferentiated blast cells were more primitive than $CFU - GEMM$. Further, using the theory of branching processes they estimated the probability of self-renewal of the blast cells. The estimate, 0.589, compared favorably with the value of 0.61 obtained by Vogel et al. (1969).

Kurnit et al. (1985) studied the *in vitro* growth of erythroid colonies from human peripheral blood mononuclear cells. Commitment in the erythroid lineage is easily detected visually by the appearance of red hemoglobin in the cells. Kurnit et al. concentrated on modeling the time until the first commitment event, and the colony size at that time. They also derived expressions for the colony size at the time when all cells were committed, and the likelihood that the colony would contain only committed cells. Two versions of a fixed generation time model were proposed; both allowed asymmetric division of a parent cell with independent fates for the two daughter cells. In the first version (Fig. 1.3e), the stem cell divides and, with probability p, each daughter cell can differentiate into an erythroid progenitor cell. In the second version (Fig. 1.3e), with the same probability p, the stem cell can differentiate

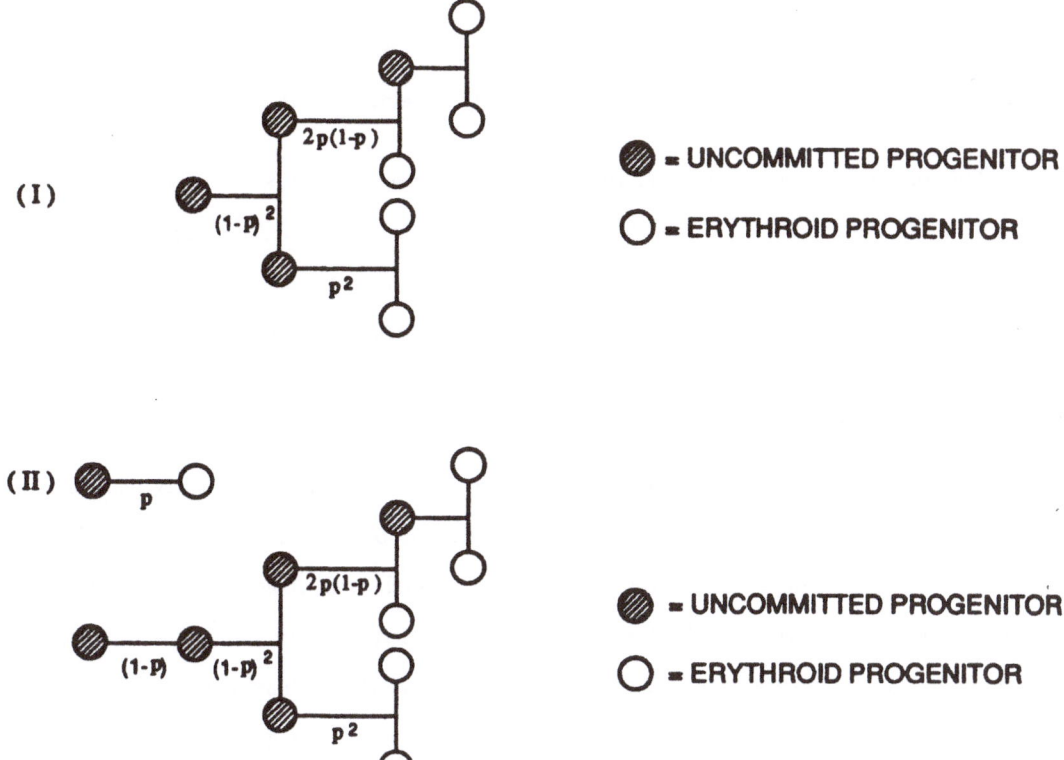

Figure 1.3 continued.   (e) Model of Kurnit et al. (1985). Single-type models of colony growth in clonal assays. (I) At each discrete time point, an uncommitted progenitor divides to give rise to two, one, or no uncommitted progenitors, with probabilities $(1-p)^2$, $2p(1-p)$, and p^2, respectively. Differentiated offspring are erythroid progenitors. (II) At the first time point, an uncommitted progenitor may commit without dividing, with probability p. From the second time point onwards, model II is identical to model I.

directly into a progenitor without dividing. Using equations from the second version to estimate p, they found that the value of p in twenty-seven experiments ranged from 0.40 to 0.63 with seven estimates less than 0.5. Kurnit et al. showed that $p < 0.5$ is the condition for a colony to grow without bound in their model. They predicted two qualitative behaviors in growing colonies: first, the colony size at first commitment increases as p decreases; and second, there is great variability in the colony size at the time of first commitment for any value of p. They also deduced, as we do later in this monograph, that major aspects of population structure can be derived from the study of relatively small colonies.

The mast cell is an exception to the rule that mature cells lack the ability to proliferate. Pharr et al. (1985) studied proliferation of mast cells using an *in vitro* clonal assay. They developed a discrete time model (see Fig. 1.3f) in which a mast cell may die or "disappear" (with probability p_0) or may divide at the end of each cell cycle into two proliferating cells (with probability p_2) or two non-proliferating cells (with probability $1 - p_0 - p_2$). A non-proliferating cell may disappear (with probability p_0) or do nothing (with probability $1 - p_0$). Colonies were initiated with one proliferative cell. Each initiated colony was assigned a value of p_2 chosen from a specified two-parameter probability distribution (the Johnson S_B distribution). Monte Carlo simulations of the model, with fixed generation times and subsequent comparison with their experimental data, led to estimates of p_0 and the two parameters that determined p_2. The model, because of the inclusion of non-proliferating cells, successfully predicted that colony size distributions can be long-tailed or even bimodal. The model also predicted that the relative number of proliferating cells should be less in small colonies than in large colonies. This prediction was tested in replating experiments, and the ratio of proliferative to non-proliferative cells in small and large colonies were judged to be significantly different.

A more general model of mast cell proliferation was studied in Nedelman et al. (1987). Mast cells were characterized as initiating, proliferating, or non-proliferating cells. A colony was assumed to be initiated by cells that have the same probabilities for offspring production as proliferating cells, but initiating cells were assumed to have a random lag-time before their first division. At the end of each cell cycle, an initiating or proliferating cell was assumed to produce two offspring, 0, 1, or 2 of which were proliferating cells, and the remainder non-proliferating cells, with probabilities p_{02}, p_{11}, and p_{20}, respectively (see Fig. 1.3g). A

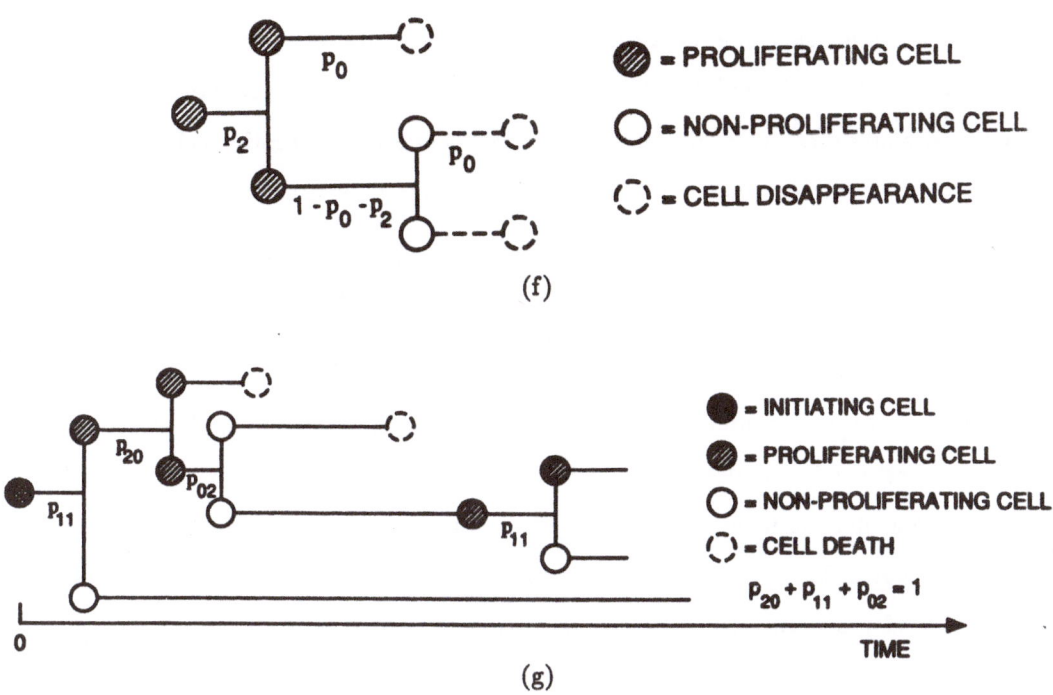

(f)

(g)

Figure 1.3 continued. Two multitype models of clonal growth of mast cells *in vitro*. (f) Model of Pharr et al. (1985). At the end of each discrete time period, a proliferating mast cell can divide to produce either two mast cells or two non-proliferating cells, with probabilities p_2 and $1 - p_0 - p_2$, respectively; or it can disappear without dividing, with probability p_0. A non-proliferating cell can also disappear without dividing, with probability p_0. (g) Model of Nedelman et al. (1987). Continuous time model. Initiating cells can, after a randomly distributed time, divide to produce two, one, or no proliferating cells, with probabilities p_{20}, p_{11}, and p_{00}, respectively. Proliferating cells undergo division after a randomly distributed time, possibly different from the time to division of an initiating cell, but the time chosen from the same probability distribution. A non-proliferating cell may come back into cycle after a randomly distributed time chosen from a different probability distribution. Death for cells other than initiators may occur before a division has occurred. Time to cell death is distributed randomly and independently of cell division times.

non-proliferating cell was assumed to become proliferative after some random time. Cell lifetime and cell division time were chosen to be random variables with distinct probability distributions. The model was fitted to experimental data on colony growth, leading to parameter estimates that suggested the amount of cell death was frequently negligible. Further, non-proliferative cells occurred at a low frequency and rarely became proliferative within the time span of the experiment. Colony size distributions were not particularly bimodal, but the occurrence of asymmetric divisions was too frequent to be ignored, as indicated by estimates of p_{11}. In one experiment out of a total of three, $p_{11} = 0.45$.

A recurrent theme in studies of cell proliferation and differentiation has been the attempt to infer properties of an unidentifiable precursor cell, such as a stem cell, from the behavior of distinguishable differentiated offspring. Theoretical approaches have tended to be "purpose-built" for a particular cell system and therefore lack general applicability. In this study, we begin to develop a framework for a coherent study of hemopoietic stem cell differentiation and proliferation.

We begin in Chapter 2 by describing in detail our model for cellular proliferation and differentiation. In Chapter 3 we characterize the dynamics of colony* growth. We first compute the total colony size as a function of time and then develop criteria for determining if a colony will cease growing. We also examine the growth of stem cells within a colony and compute the probability that all of the stem cells in a colony will disappear by differentiating into progenitor cells. In Chapter 4, for those colonies that reach completion, we compute the distribution of the time it takes to reach completion and the size at completion. In Chapter 5 we describe colonies that grow without bound by computing the asymptotic proportions of cell types within the colony. As the parameters of our model are changed, it is seen that a threshold exists between colonies completing growth in finite time and colonies having a positive probability of growing without bound: Chapter 6 is devoted to the study of models balanced at this threshold. In the theory of branching processes this is known as the study

---

* We will often use the word colony to mean either a colony or a cluster, when it is not important to distinguish between the two different sized groups.

of "critical" processes. In Chapter 7 we use the theory developed in Chapters 2-6 to make numerical predictions about colony growth in culture and to compare in a qualitative way our theoretical results with experimental observations. In Chapter 8 we summarize our model and its results and provide a critical discussion of the model's usefulness in answering current biological questions. Readers wishing to omit the mathematical details should read the model description in Chapter 2 and then skip to Chapter 7 where the results are presented. Some of these results have been presented without proof in Macken et al. (1986).

Chapter 2

A Multitype Branching Process Model

The model we develop is one of the simplest nontrivial models of cell growth in culture. We assume that there are three general classes of cells, which we label stem cells (S), macrophage progenitors (M), and end cells (E). Discrimination among the classes is on the basis of function of the member cells since typically the cell types cannot be distinguished morphologically. *Stem cells* are described as having the capacity for infinite self-renewal. By this, we mean that a stem cell can divide an infinite number of times, and at each division its offspring can be an exact replica of itself. A daughter cell that has differentiated from the parent stem cell is called a *progenitor*. In contrast to stem cells, progenitors do not have the capacity for infinite self-renewal. Eventually, a progenitor *must* produce a daughter cell that differs significantly from itself: this process is called *terminal differentiation*. The fully mature offspring cannot divide and are called *end cells*. Thus the three cell types of the model form a natural progression according to increasing maturity and consequent decreasing proliferative potential; $S \rightarrow M \rightarrow E$.

Although we have loosely described a pool of macrophage progenitors, M, each with identical branching properties, progenitors are not in fact a uniform population of cells. In the case of macrophages, progenitors include monoblasts, promonocytes, and monocytes. A more detailed model might break down the progenitor compartment into subpopulations or even treat it as a distributed compartment with progenitor cells described by a continuous "maturity" variable (cf. Rubinow and Lebowitz, 1975).

Our model does not directly specify the difference between the self-renewal of a stem cell and the production of a like daughter cell by a progenitor. The capacity for infinite self-renewal will be represented indirectly via the selection of values for parameters of the model. More precise discrimination between the two types of cell-division could be built into a model by defining the progenitor class to be heterogeneous and then allowing the offspring of a progenitor to belong to the same progenitor class, yet differ from its parent, for example, by introducing a maturity variable. Further, there is no requirement in our model that the cells which we call stem cells, S, be the unique hemopoietic pluripotent stem cell described in the Preface: All that we require is that "stem" cells have the capacity for self-renewal and can differentiate into progenitor cells which have less proliferative capacity. In applications, cells, such as the CFU-S described by Till and McCulloch (1961) and discussed in the Preface, or the CFU-GM which gives rise to the cells of the granulocyte and macrophage lineages, could fulfill the role of stem cells in our model if their capacity for self-renewal were sufficiently great (although not necessarily greater than 0.5).

In modeling cell proliferation and differentiation, time can be increased in discrete units or continuously. Here, as in the models of Till et al. (1964), Vogel et al. (1969), and others (see Chapter 1), we pursue the discrete time approach and assume that stem cells and macrophage progenitors have an equal and unchanging cell cycle length, which becomes the time unit in our model. This is clearly an approximation since there is some dispersion about the mean generation time. However, this dispersion time is short compared to the generation time (cf. Stewart, 1980). The use of continuous-time models would be required to model accurately cell populations in which there is a large variance in the distribution of cell cycle times.

To further specify the model, we assume the following: at each time point, every S and M cell, independently of other cells in the colony, divides and produces two viable daughter cells; the possibility of an S or M cell dying is negligible and chosen to be zero (cf. Nedelman et al., 1987); and each daughter cell, independently of the other, may differentiate into the next cell type. Of course, this ideal situation may not be attained in practice. For example, after nuclear division a cell may not undergo cytokinesis, or a cell which is not fully differentiated may slip out of cell cycle. Cells may also die in culture. Here we assume that the likelihood of these occurrences is sufficiently low under the experimental conditions of interest that they can be ignored.

The assumption that the fates of daughter cells are independent requires elaboration. There are conflicting views on whether cells undergo *symmetric division* (in which daughter cells are necessarily of the same type as each other, although possibly of a different type than their parent) or *asymmetric division* (in which daughter cells may differ in type from each other). We have chosen to assume that divisions may be asymmetric and refer to the following two reports as examples in support of our view. Metcalf (1980) carried out a definitive series of clone-transfer experiments in which the two daughter cells produced by one division of a single bipotential granulocyte-macrophage progenitor were separately transferred into identical culture conditions. Some pairs of daughter cells gave rise to colonies of different composition, thus demonstrating asymmetric division of the granulocyte-macrophage progenitor. Suda et al. (1984) used similar clone-transfer experiments to provide evidence for asymmetric differentiation of a multipotential blast cell. Further, the family tree shown in Fig. 1.2 illustrates asymmetric cell division in a single lineage, and the work of Nedelman et al. (1987), see Fig. 1.3g, indicates the occurrence of asymmetric division in the mast cell lineage.

The assumption of independence of the fates of daughter cells is in fact stronger than the assumption of asymmetric division. Whereas asymmetry allows for the correlation between fates of daughter cells to be less than one, independence of fates implies this correlation must be zero. Further details on the implications of the assumption of independence will follow in Section 2.1.

Assuming that the differentiation process is stochastic, we model the growth and differentiation of a single cell into a colony containing possibly more than one type of cell by a multitype branching process. As described by Fig. 2.1, the fundamental behavior of each cell type is characterized by the following probabilities:

$$p_s = Pr(\text{daughter of a cell of type } S \text{ is of type } S) \quad ,$$
$$p_m = Pr(\text{daughter of a cell of type } M \text{ is of type } M) \quad .$$

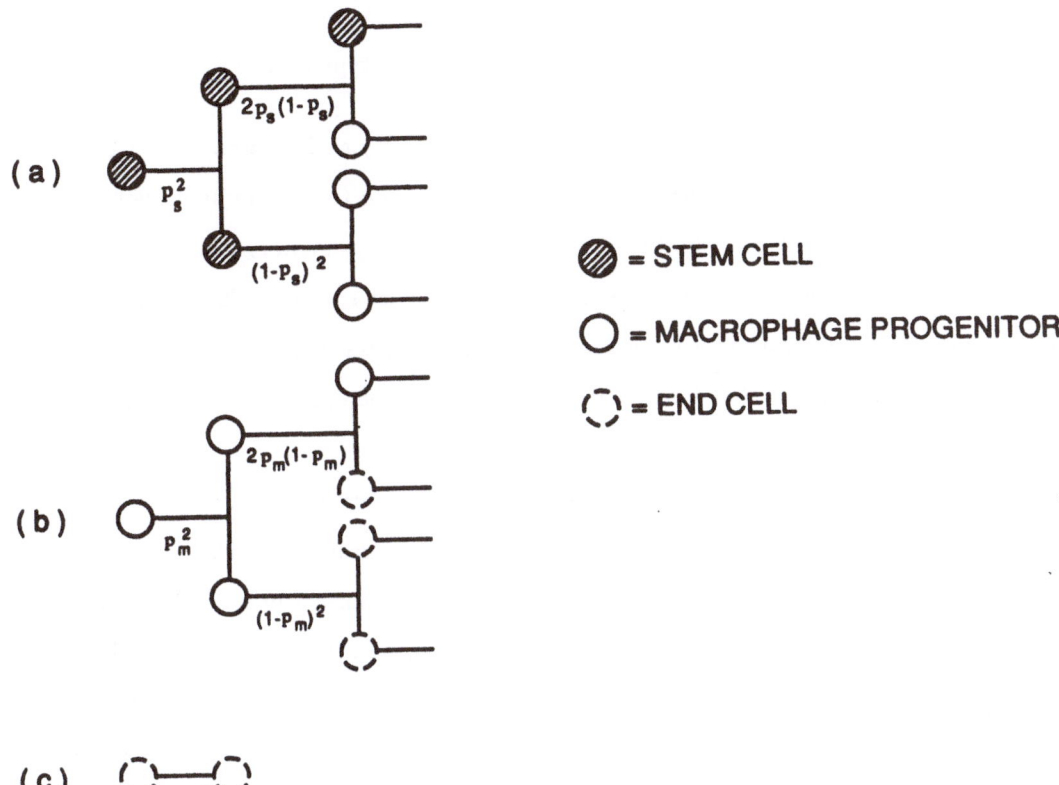

Figure 2.1   A multitype branching model of macrophage differentiation. The model includes stem cells, macrophage progenitors, and end cells (fully functional, non-dividing macrophages). Cells of any type can initiate a colony [(a)-(c)]. (a) Stem cells produce two, one, or no stem cell progeny, with probabilities $p_s^2, 2p_s(1 - p_s)$ and $(1 - p_s)^2$, respectively. (b) Macrophage progenitors produce two, one, or no macrophage progenitor offspring, with probabilities $p_m^2, 2p_m(1 - p_m)$ and $(1 - p_m)^2$, respectively. A macrophage progenitor offspring of a stem cell acts as the initiator of a subcolony with the branching structure illustrated in (b). (c) End cells do not die nor do they divide. Thus they contribute one end cell to the next generation, with probability 1.

To model the non–proliferation of end cells, we require that at each increment in time an end cell produce exactly one end cell. Thus

$$p_e = Pr(\text{daughter of a cell of type } E \text{ is of type } E) \equiv 1 \quad .$$

The probabilities p_s, p_m, and p_e are known as *branching probabilities*. Although it may appear that we are requiring differentiation and cell division to be simultaneous events, in fact this is not the case: Our definition of the probabilities p_s, p_m, and p_e simply requires that differentiation occur sometime between successive cell divisions. Note that, as defined, the probability of a daughter being of the same type as its parent is constant, independent of time. Clearly, the assumption of time-homogeneous branching probabilities can be challenged. One might argue that a cell in a given compartment "ages" according to the number of generations since an ancestral cell in its lineage first entered the compartment and that therefore the branching probabilities should decrease with age. Although our formulation may be idealistic, to allow p_s, p_m, and p_e to vary with time would introduce major mathematical complications at a stage in which we are seeking broad behavioral characterizations from our model. It is for similar reasons that we have chosen to ignore cell death, although the death of end cells can be easily modeled by choosing $p_e < 1$. Fortuitously, it appears that in experiments carried out for a period of a few weeks in good culture conditions cell death is negligible (Stewart, personal communication; Nedelman et al., 1987).

2.1 Probability Generating Functions

The outcome of one cell cycle (i.e., one "generation," in branching process terminology) can be summarized in the multivariate probability generating function of the offspring distributions. We begin by defining the probability generating function for offspring of an S cell. Let

$$f_s(\boldsymbol{\theta}) = \sum_{i,j \geq 0} p_s(i,j)\theta_s^i \theta_m^j \quad , \tag{2.1}$$

where

$$p_s(i, j) = Pr(\text{parent of type } S \text{ produces } i \text{ daughters of type } S$$

$$\text{and } j \text{ daughters of type } M \text{ in a single division}) \quad , \tag{2.2}$$

and

$$\boldsymbol{\theta} = (\theta_s, \theta_m, \theta_e)$$

is a dummy vector variable satisfying the constraint $\|\boldsymbol{\theta}\| \leq 1$. For the cells of interest, $p_s(i, j) = 0$ unless $i + j = 2$ and $i, j \geq 0$.

As discussed in the previous section, we assume that daughter cells have independent fates. Thus the outcome of an S cell division is equivalent to the outcome of a binomial experiment consisting of two trials, each with probability p_s of success (if we define self-replication as a success). Therefore,

$$p_s(i, j) = \binom{2}{i} p_s^i (1 - p_s)^j \quad ,$$

where $i + j = 2$, from which we obtain

$$f_s(\boldsymbol{\theta}) = \sum_{i=0}^{2} \binom{2}{i} p_s^i (1 - p_s)^{2-i} \theta_s^i \theta_m^{2-i} = [p_s \theta_s + (1 - p_s)\theta_m]^2 \quad . \tag{2.3}$$

The assumption of independence of fates leads to the probability of heterogeneous offspring $[2p_s(1 - p_s)]$ being greater than the probability of one of the homogeneous alternatives $[p_s^2$ or $(1 - p_s)^2]$. In some systems, such as the p, q, r model of Vogel et al. (1969) for the development of erythroblast colonies, correlations may exist between the fates of daughter cells. In such cases the more powerful formula (2.1) would be needed.

In a fashion similar to that used above, we derive the probability generating function for offspring of a type M cell. Recall that daughters may now be of type M or E. Thus

$$f_m(\boldsymbol{\theta}) = \sum_{i=0}^{2} \binom{2}{i} p_m^i (1 - p_m)^{2-i} \theta_m^i \theta_e^{2-i} = [p_m \theta_m + (1 - p_m)\theta_e]^2 \quad . \tag{2.4}$$

Finally, the degenerate case of the probability generating function for offspring of a type E cell is just

$$f_e(\boldsymbol{\theta}) = p_e \theta_e = \theta_e \quad . \tag{2.5}$$

Equations (2.3)-(2.5) can be summarized in vector form by defining

$$\mathbf{f}(\boldsymbol{\theta}) = (f_s(\boldsymbol{\theta}), f_m(\boldsymbol{\theta}), f_e(\boldsymbol{\theta})) \quad . \tag{2.6}$$

We now consider a cell of a given type, plated at time zero, and ask for a probabilistic description of the composition of the colony that arises from it after n generations, i.e., at time n. The method we shall use depends upon an important result in branching processes. First, it is necessary to introduce some notation. Let

$Z_{\alpha n}$ = number of cells of type α in the colony after n generations, $\alpha = S, M, E$,

and write

$$\mathbf{Z}_n = (Z_{sn}, Z_{mn}, Z_{en}) \quad .$$

The cell that gives rise to the colony constitutes the zeroth generation. We misuse notation and write

$$\mathbf{Z}_0 = S \quad , \quad \text{or} \quad \mathbf{Z}_0 = M \quad , \quad \text{or} \quad \mathbf{Z}_0 = E$$

for the unit vectors corresponding to each of the possible zeroth generations. (In the sequel when $\mathbf{Z}_0$ is used in matrix operations, the appropriate unit vector will need to be substituted.) Then, Eq. (2.2) can be written as

$$p_s(i, j) = Pr((\mathbf{Z}_1 = (i, j, 0)|\mathbf{Z}_0 = S) \quad . \tag{2.7a}$$

The equivalent expression for parents of type M is

$$p_m(i, j) = Pr((\mathbf{Z}_1 = (0, i, j)|\mathbf{Z}_0 = M) \quad . \tag{2.7b}$$

Definitions (2.3) and (2.4) are extended to apply to n generations as follows.

Define

$$f_\alpha^{(n)}(\boldsymbol{\theta}) = \sum_{i \geq 0} \sum_{j \geq 0} \sum_{k \geq 0} p_\alpha^{(n)}(i, j, k)\theta_s^i \theta_m^j \theta_e^k \quad , \tag{2.8a}$$

where

$$p_\alpha^{(n)}(i,j,k) = Pr(Z_n = (i,j,k)|Z_0 = \alpha) \quad . \tag{2.8b}$$

An important result in branching processes is (cf. Karlin and Taylor, 1975)

$$\mathbf{f}^{(n)}(\boldsymbol{\theta}) = \mathbf{f}^{(1)}(\mathbf{f}^{(n-1)}(\boldsymbol{\theta})) = \mathbf{f}^{(n-1)}(\mathbf{f}^{(1)}(\boldsymbol{\theta})) \quad . \tag{2.9}$$

By comparing Eq. (2.8) with $n = 1$, with Eq. (2.1) after substituting Eq. (2.7), we obtain

$$f_\alpha^{(1)}(\boldsymbol{\theta}) = f_\alpha(\boldsymbol{\theta}) \quad , \quad \alpha = S, \ M, \ E \quad . \tag{2.10}$$

By iterating Eq. (2.9) and using Eq. (2.10), we obtain the result that provides the foundation of our modeling, namely

$$\mathbf{f}^{(n)}(\boldsymbol{\theta}) = \underbrace{\mathbf{f}(\mathbf{f}(\mathbf{f}\ldots(\boldsymbol{\theta})\ldots))}_{n \text{ times}} \equiv \mathbf{f}^n(\boldsymbol{\theta}) \quad , \tag{2.11}$$

where $\mathbf{f}^n$ denotes the nth iterate of $\mathbf{f}$ and $\mathbf{f}^{(n)}$ is the generating function for the nth generation. Note that each element of $\mathbf{f}^n(\boldsymbol{\theta})$ will be a function (albeit complicated) of the branching probabilities p_s, p_m, and p_e, and the dummy variable θ. In principle, therefore, we are afforded a means of calculating the probability distribution for colony composition after n generations as a function of p_s, p_m, and p_e: We simply iterate the probability generating function n times as described in Eq. (2.11), then extract the coefficients of $\theta_s^i \theta_m^j \theta_e^k$. By Eq. (2.8) these coefficients give the distribution of cell types in the colony Z_n. Unfortunately, in practice, the process of iterating the probability generating function leads to extremely cumbersome formulae, which restricts the general applicability of this approach.

2.2   Moments

We can use iterative procedures for calculating the first two moments of Z_n as a function of n and the branching probabilities. It can be readily shown (cf. Karlin and Taylor,

1975; Appendix A) that the mean population composition after n generations, given that the population composition at the zeroth generation is $\mathbf{Z_0}$, is given by

$$E[\mathbf{Z}_n|\mathbf{Z}_0] = \mathbf{Z}_0\mathbf{M}^n \quad , \quad n \geq 1 \tag{2.12}$$

where

$$\mathbf{M} = \begin{pmatrix} 2p_s & 2(1-p_s) & 0 \\ 0 & 2p_m & 2(1-p_m) \\ 0 & 0 & 1 \end{pmatrix} , \tag{2.13}$$

and $E(\cdot)$ denotes the mean. The (i, j)th entry of the matrix of means, $\mathbf{M}$, gives the mean number of offspring of type j contributed by a parent of type i in one generation, with $i, j = S, M, E$.

Using Eqs. (2.12) and (2.13), we show in Appendix A that

(i) $E[\mathbf{Z}_n|\mathbf{Z}_0 = S] =$

$$
\begin{cases}
\left((2p_s)^n, \; \dfrac{(1-p_s)}{(p_s-p_m)}[(2p_s)^n - (2p_m)^n], \dfrac{2(1-p_s)(1-p_m)}{(p_s-p_m)} \times \right. \\
\quad \left. \left[\dfrac{(2p_s)^n}{(2p_s-1)} - \dfrac{(2p_m)^n}{(2p_m-1)} \right] + \dfrac{4(1-p_s)(1-p_m)}{(2p_s-1)(2p_m-1)} \right), & p_s \neq p_m; p_s, p_m \neq 0.5 \\[4mm]
\left(1, \; \dfrac{1-(2p_m)^n}{1-2p_m}, \; \dfrac{2(1-p_m)[n(1-2p_s)-1+(2p_s)^n]}{(1-2p_m)^2} \right), & p_s = 0.5, p_m \neq 0.5 \\[4mm]
\left((2p_s)^n, \dfrac{2(1-p_s)[1-(2p_s)^n]}{1-2p_s} , \right. \\
\quad \left. \dfrac{2(1-p_s)[n(1-2p_s)-1+(2p_s)^n]}{(1-2p_s)^2} \right), & p_m = 0.5, p_s \neq 0.5 \\[4mm]
\left((2p)^n, \; 2n(1-p)(2p)^{n-1}, \right. \\
\quad \left. \dfrac{4(1-p)^2[1-(2p)^n]}{(1-2p)^2} - \dfrac{4n(1-p)^2(2p)^{n-1}}{1-2p} \right), & p_s = p_m = p \neq 0.5 \\[4mm]
\left(1, \; n, \; \dfrac{n(n-1)}{2} \right), & p_s = p_m = 0.5 \; ;
\end{cases}
$$

$$\tag{2.14a}$$

(ii) $E[\mathbf{Z}_n|\mathbf{Z}_0 = M] =$

$$
\begin{cases}
\left(0,\ (2p_m)^n,\ \dfrac{2(1-p_m)[1-(2p_m)^n]}{1-2p_m}\right), & p_m \neq 0.5 \\[4mm]
(0,1,n), & p_m = 0.5 \quad ;
\end{cases}
$$

(2.14b)

(iii) $E[\mathbf{Z}_n|\mathbf{Z}_0 = E] = (0,0,1)$.   (2.14c)

The variance in the population composition can also be found. We show in Appendix A that the variance obeys the following recursion.

$$
var[\mathbf{Z}_n|\mathbf{Z}_0] = E[var(\mathbf{Z}_n|\mathbf{Z}_{n-1})] + \mathbf{M}'\, var[\mathbf{Z}_{n-1}|\mathbf{Z}_0]\mathbf{M} \quad , \quad n \geq 1 \ , \qquad (2.15)
$$

where $\mathbf{M}'$ is the transpose of $\mathbf{M}$.

The solution to recursion relation (2.15) is a symmetric 3 x 3 matrix,

$$
var[\mathbf{Z}_n|\mathbf{Z}_0] =
\begin{bmatrix}
var[Z_{sn}|\mathbf{Z}_0] & cov[Z_{sn}, Z_{mn}|\mathbf{Z}_0] & cov[Z_{sn}, Z_{en}|\mathbf{Z}_0] \\
cov[Z_{sn}, Z_{mn}|\mathbf{Z}_0] & var[Z_{mn}|\mathbf{Z}_0] & cov[Z_{mn}, Z_{en}|\mathbf{Z}_0] \\
cov[Z_{sn}, Z_{en}|\mathbf{Z}_0] & cov[Z_{mn}, Z_{en}|\mathbf{Z}_0] & var[Z_{en}|\mathbf{Z}_0]
\end{bmatrix}
\quad , \quad n \geq 1 \ ,
$$

(2.16)

in which each element is a function of n, p_s, and p_m. If we introduce the notation,

$$
var[\mathbf{Z}_n|\mathbf{Z}_0] =
\begin{pmatrix}
u_n & v_n & w_n \\
v_n & x_n & y_n \\
w_n & y_n & z_n
\end{pmatrix} \ ,
$$

Eq. (2.16) can be written as

$$
\begin{bmatrix} u_{n+1} \\ v_{n+1} \\ w_{n+1} \\ x_{n+1} \\ y_{n+1} \\ z_{n+1} \end{bmatrix} = \begin{bmatrix} \mu_{1n} \\ -\mu_{1n} \\ 0 \\ \mu_{1n} + \mu_{2n} \\ -\mu_{2n} \\ \mu_{2n} \end{bmatrix} +
$$

$$
\begin{bmatrix} 4p_s^2 & 0 & 0 & 0 & 0 & 0 \\ 4p_s(1-p_s) & 4p_sp_m & 0 & 0 & 0 & 0 \\ 0 & 4p_s(1-p_m) & 2p_s & 0 & 0 & 0 \\ 4(1-p_s)^2 & 8p_m(1-p_s) & 0 & 4p_m^2 & 0 & 0 \\ 0 & 4(1-p_s)(1-p_m) & 2(1-p_s) & 4p_m(1-p_m) & 2p_m & 0 \\ 0 & 0 & 0 & 4(1-p_m)^2 & 4(1-p_m) & 1 \end{bmatrix} \begin{bmatrix} u_n \\ v_n \\ w_n \\ x_n \\ y_n \\ z_n \end{bmatrix} ,
$$

$$(2.17)$$

where

$$\mu_{1n} = 2p_s(1-p_s)E[Z_{sn}|Z_0 = \alpha] \quad,$$

$$\mu_{2n} = 2p_m(1-p_m)E[Z_{mn}|Z_0 = \alpha] \quad,$$

and

$$u_0 = v_0 = w_0 = x_0 = y_0 = z_0 = 0 \quad.$$

For the case $p_s = p_m = 0.5$, Eq. (2.17) is solved explicitly in Appendix A. For other values of p_s and p_m, an explicit solution seems too cumbersome to be useful. In this case numerical solutions are easily obtained, and such numerical solutions are used later in the text.

The mean and variance of Z_n provide convenient summaries of some of the major features of the full distribution function for Z_n obtained by the generating function approach above. However, caution must be exercised in their use because the distribution of Z_n is not "nicely" behaved, i.e., higher order moments do not decrease to zero with increasing n.

2.3 Modeling Unknown Types of Colony Initiating Cells

One further modeling step is necessary before we have a theory that can be compared with experiment. This step must take account of the fact that experimental observations are generally made on the growth of colonies initiated by plating out cells from the bone marrow, which are of mixed and unknown cell types. The probability distribution for the composition of an entire plate of cells can be expressed as a function of p_s, p_m, and p_e by writing

$$Pr(Z_n = (i, j, k)) = \sum_\alpha \eta_\alpha Pr(Z_n = (i, j, k)|Z_0 = \alpha) \quad , \tag{2.18}$$

where η_s, η_m, η_e are the proportions of cells plated at time 0 of each of the three types. The mean of Z_n for plates of initiating cells of mixed types is

$$E[Z_n] = \sum_\alpha \eta_\alpha E[Z_n|Z_0 = \alpha] \quad . \tag{2.19}$$

To obtain the corresponding variance of Z_n, we appeal to a standard result in statistical theory (cf. Mood, Graybill, and Boes, 1974) to show that

$$var[Z_n] = E_{Z_0}(var[Z_n|Z_0]) + var_{Z_0}(E[Z_n|Z_0]) \quad , \tag{2.20}$$

where the subscript of Z_0 on the expectation and variance operators indicates that the operation is applied with respect to the distribution of Z_0. From the theory of linear models (cf. Searle, 1971) we obtain

$$var_{Z_0}(E[Z_n|Z_0]) = var_{Z_0}(Z_0 M^n) = (M^n)'var[Z_0]M^n \quad , \tag{2.21}$$

where M is given in Eq. (2.13) and "′" indicates transpose. Finally

$$Z_0 = \begin{cases} (1, 0, 0) & \text{with probability } \eta_s \\ (0, 1, 0) & \text{with probability } \eta_m \\ (0, 0, 1) & \text{with probability } \eta_e \end{cases} ,$$

and hence

$$var[\mathbf{Z}_0] \equiv \begin{bmatrix} var(Z_{s0}) & cov(Z_{s0}, Z_{m0}) & cov(Z_{s0}, Z_{e0}) \\ cov(Z_{s0}, Z_{m0}) & var(Z_{m0}) & cov(Z_{m0}, Z_{e0}) \\ cov(Z_{s0}, Z_{e0}) & cov(Z_{m0}, Z_{e0}) & var(Z_{e0}) \end{bmatrix}$$

$$= \begin{bmatrix} \eta_s(1 - \eta_s) & -\eta_s\eta_m & -\eta_s\eta_e \\ -\eta_s\eta_m & \eta_m(1 - \eta_m) & -\eta_m\eta_e \\ -\eta_s\eta_e & -\eta_m\eta_e & \eta_e(1 - \eta_e) \end{bmatrix} . \qquad (2.22)$$

By substituting Eq. (2.21) into Eq. (2.20), we obtain

$$var[\mathbf{Z}_n] = \sum_{\alpha} \eta_\alpha var[\mathbf{Z}_n | \mathbf{Z}_0 = \alpha] + (\mathbf{M}^n)' var[\mathbf{Z}_0]\mathbf{M}^n \quad , \qquad (2.23)$$

where $var[\mathbf{Z}_0]$ is given by Eq. (2.22).

Equation (2.20) can be interpreted in an analysis of variance style as follows. The total variability observed among plated colonies $(var[\mathbf{Z}_n])$ is composed of two separable components. The first component $(E_{\mathbf{Z}_0}(var[\mathbf{Z}_n | \mathbf{Z}_0)]))$ measures the average variability *within* a colony, where the average is taken over all possible types of initiating cells. The "within" variability is due to the stochastic nature of the proliferative process. The second component $(var_{\mathbf{Z}_0}(E[\mathbf{Z}_n | \mathbf{Z}_0]))$ measures the variability *among* the *average* behavior of colonies due to different types of initiating cells.

Chapter 3

Characterization of Colony Growth with Time

In the previous chapter, we alluded to the problem of iteration of the generating functions: Essentially it is computationally infeasible to evaluate Eq. (2.11) for large n. Hence, we cannot obtain the detailed probability distribution for colony composition with time, $Pr(\mathbf{Z}_n = (i, j, k)|\mathbf{Z}_0)$. However, in this and the following three chapters we characterize other aspects of colony growth in finite time and obtain results in the limit $n \to \infty$. Because of the recursive nature of many of the expressions, explicit analytical results are not obtainable except in rather trivial special cases. The implications of the expressions will therefore not become apparent until Chapter 7 where we provide numerical examples for selected values of the branching probabilities.

3.1 Total Colony Size after n Generations

The first quantity of interest is the total colony size, irrespective of composition by cell-type. We obtain a recursion that allows us to calculate the probability distribution for the colony size after a finite number of generations. The essence of the recursion is to consider, for example, a parent of type S and ask for the probability that after n generations the colony arising from this parent has size k. After the first cell division, two offspring are generated, each of which may be either an S or an M cell. These offspring are assumed to divide independently of each other, and thus each can be considered to be the parent of a subcolony. After $n - 1$ additional generations, we require that the combined sizes of these two subcolonies be k. To formulate this argument mathematically, we let

$$q_\alpha^{(n)}(k) = Pr(\text{parent of type } \alpha \text{ produces a colony of size } k \text{ in } n \text{ generations}) \quad ,$$

$$\alpha = S, M \; ; \quad k \geq 0 \; ; \quad n \geq 0 \quad .$$

After one cell division, with probabilities p_s^2, $2p_s(1 - p_s)$, and $(1 - p_s)^2$, the two offspring of a stem cell are both S cells, one is an S cell and one is an M cell, and both are M cells, respectively. By conditioning on the first cell division (first generation), we obtain for $n \geq 1$,

$$
q_s^{(n)}(k) = p_s^2 \left[\sum_{j=0}^{k} q_s^{(n-1)}(j) q_s^{(n-1)}(k - j) \right]
$$

$$
+ 2p_s(1 - p_s) \left[\sum_{j=0}^{k} q_s^{(n-1)}(j) q_m^{(n-1)}(k - j) \right] \tag{3.1a}
$$

$$
+ (1 - p_s)^2 \left[\sum_{j=0}^{k} q_m^{(n-1)}(j) q_m^{(n-1)}(k - j) \right] .
$$

If the parent cell is a macrophage, then after one generation with probabilities p_m^2, $2p_m(1 - p_m)$, and $(1 - p_m)^2$, the offspring are both M cells, one is an M cell and one is an E cell, and both are E cells, respectively. In this case the recursion becomes

$$
q_m^{(n)}(k) = p_m^2 \left[\sum_{j=0}^{k} q_m^{(n-1)}(j) q_m^{(n-1)}(k - j) \right]
$$

$$
+ 2p_m(1 - p_m) q_m^{(n-1)}(k - 1) + (1 - p_m)^2 \delta_{k,2} , \tag{3.1b}
$$

where $\delta_{k,2}$ is the Kronecker delta, which takes on the value 1 if $k = 2$ and 0 otherwise. The $\delta_{k,2}$ term accumulates end cells that can no longer divide.

Because colonies start with a single cell and we are not allowing cell death,

$$
q_\alpha^{(n)}(0) = 0 \quad , \quad n \geq 0 \quad ; \quad \alpha = S, M \tag{3.2a}
$$

and

$$
q_\alpha^{(0)}(k) = \delta_{k,1} \quad , \quad u - S, M \quad ; \quad h \geq 0 \quad . \tag{3.2b}
$$

Further,

$$q_\alpha^{(n)}(k) = 0 \quad , \quad \alpha = S, M \quad ; \quad k > 2^n \quad ; \quad n \geq 1 \quad , \tag{3.2c}$$

since after n generations a maximum of 2^n cells can be produced.

Using Eq. (3.2b) to commence a recursion, we find

$$q_m^{(1)}(1) = 0 \quad , \quad q_m^{(1)}(2) = 1 \quad ,$$

$$q_s^{(1)}(1) = 0 \quad , \quad q_s^{(1)}(2) = 1 \quad ;$$

i.e., both a macrophage and a stem cell have exactly two offspring after one generation.

At the second generation

$$q_m^{(2)}(1) = 0 \quad , \quad q_m^{(2)}(2) = (1 - p_m)^2 \quad ,$$

$$q_m^{(2)}(3) = 2p_m(1 - p_m) \quad , \quad q_m^{(2)}(4) = p_m^2 \quad .$$

Thus, if the macrophage gives rise to two end cells, an event that occurs with probability $(1 - p_m)^2$, it will have only two offspring in the second generation. If one offspring is an end cell and one is a macrophage, an event that occurs with probability $2p_m(1 - p_m)$, then in the second generation there will be three cells at least one of which is an end cell. Lastly, if both offspring are macrophages, an event with probability p_m^2, then in the second generation there will be four cells. For a stem cell parent, in the second generation

$$q_s^{(2)}(1) = q_s^{(2)}(2) = q_s^{(2)}(3) = 0 \quad , \quad q_s^{(2)}(4) = 1 \quad ;$$

i.e., a stem cell must have four offspring by the second generation.

One can continue this recursion algebraically, but it becomes extremely complicated as n increases. However, once values are assigned to p_s and p_m, the recursion can be solved numerically. In Chapter 7 we provide some illustrative results.

Another approach to describing the random total colony size after n generations is via its moments. We use the results for $E[\mathbf{Z}_n|\mathbf{Z}_0]$ and $var[\mathbf{Z}_n|\mathbf{Z}_0]$ from Appendix A in the following expressions:

$$E[Z_{sn} + Z_{mn} + Z_{en}|\mathbf{Z}_0] = E[Z_{sn}|\mathbf{Z}_0] + E[Z_{mn}|\mathbf{Z}_0] + E[Z_{en}|\mathbf{Z}_0] \qquad (3.3a)$$

and

$$var[Z_{sn} + Z_{mn} + Z_{en}|\mathbf{Z}_0] = var[Z_{sn}|\mathbf{Z}_0] + 2cov[Z_{sn}, Z_{mn}|\mathbf{Z}_0]$$
$$2cov[Z_{sn}, Z_{en}|\mathbf{Z}_0] + var[Z_{mn}|\mathbf{Z}_0] + 2cov[Z_{mn}, Z_{en}|\mathbf{Z}_0] + var[Z_{en}|\mathbf{Z}_0] \quad .$$
$$(3.3b)$$

Both the mean and the variance of the total colony size increases with n, with the variance being much greater than the mean for all n. Unfortunately, because of the complexity of the probability distribution of total colony size, it is not easy to determine the proportion of colonies that are expected to fall within the bounds of $mean \pm \sqrt{variance}$. An additional difficulty in interpreting these bounds is due to the fact that the proportion of non-proliferating colonies increases with n. If we condition on a colony being proliferative, the conditional mean colony size will be substantially greater than the unconditional mean obtained by the method of Appendix A. On the other hand, the increasing proportion of non-proliferative colonies will have the effect of increasing the unconditional variance above that which would be obtained if we conditioned on the colony being proliferative. Results for conditional moments of the process are not obtained here; the unconditional moments are, however, quite appropriate for small values of n.

After growth has been allowed to continue for a long time, it becomes likely that colonies of small size will consist entirely of end cells. We therefore turn next to techniques for discriminating between those colonies that contain only end cells, and have, therefore, completed their growth, and those colonies that are still growing.

3.2   Probability of Completion of Growth

Experiments indicate that with increasing time colonies tend to dichotomize into small, non-growing colonies or very large, growing colonies. We compare this observation with the general behavior of our model by noting that there is only one absorbing state in our three–compartment system, namely, the state in which all cells belong to the E compartment. Consequently, the system either reaches a state, in a finite time, in which all cells are a member of the E compartment, or it never reaches that state. The implication in terms of colony growth is that a colony either completes its growth in a finite number of generations or continues to grow without bound. We can calculate the probability of completion of growth as a function of the branching probabilities p_s, p_m, and p_e.

To do so, we first make an artificial adjustment to our model that allows us to consider a simpler equivalent problem. Suppose that M cells can produce offspring that are either of their own type or die. Then the system consisting of only S and M cells constitutes a two-type branching process. Completion of growth in the original model corresponds to extinction (i.e., death of all cells) in the modified model. The probability generating function for offspring of an M cell must be correspondingly adjusted. The outcome of an M cell division is now a binomial experiment with two independent trials where each trial results in success (offspring of type M) with probability p_m or in failure (offspring dies) with probability $1 - p_m$.

If we define the probability generating function of the modified process by

$$\mathbf{g}(\boldsymbol{\phi}) = (g_s(\boldsymbol{\phi}), g_m(\boldsymbol{\phi})) \quad ,$$

where

$$\boldsymbol{\phi} = (\phi_s, \phi_m) \quad ,$$

then

$$g_m(\boldsymbol{\phi}) = [p_m \phi_m + (1 - p_m)]^2 \quad . \tag{3.4a}$$

The offspring distribution of an S cell is unchanged, and so from Eq. (2.3),

$$g_s(\boldsymbol{\phi}) = [p_s \phi_s + (1 - p_s)\phi_m]^2 \quad . \tag{3.4b}$$

3.2.1   Completion of growth in finite time

In the modified process, define

$$\mathbf{Y}_n = (Y_{sn}, Y_{mn}) \ ,$$

where

$$Y_{\alpha n} = \text{number of cells of type } \alpha \text{ in the colony at time } n \geq 0 \ , \ \alpha = S, M \ \ .$$

Then, as in Eq. (2.8), we have

$$g_\alpha^{(n)}(\boldsymbol{\phi}) = \sum_{i \geq 0} \sum_{j \geq 0} Pr(\mathbf{Y}_n = (i,j) | \mathbf{Y}_0 = \alpha) \phi_s^i \phi_m^j \ , \tag{3.5}$$

which, by Eq. (2.11), can be calculated as

$$g_\alpha^{(n)}(\boldsymbol{\phi}) = g_\alpha \underbrace{(\mathbf{g}(\mathbf{g}(\ldots(\boldsymbol{\phi})\ldots)))}_{(n-1) \ times}$$

$$= g_\alpha(\mathbf{g}^{n-1}(\boldsymbol{\phi})) \quad .$$

By setting $\boldsymbol{\phi} = \mathbf{0}$ in Eq. (3.5), we obtain

$$g_\alpha^{(n)}(\mathbf{0}) = Pr(\mathbf{Y}_n = \mathbf{0} | \mathbf{Y}_0 = \alpha)$$

$$= g_\alpha(\mathbf{g}^{n-1}(\boldsymbol{\phi}))|_{\boldsymbol{\phi}=0} \quad , \cdot \tag{3.6}$$

which is the probability that a colony has become extinct (i.e., reached completion in the original process) by n generations. Define

$$\pi_{\alpha n} - Pr(\mathbf{Y}_n = \mathbf{0} | \mathbf{Y}_0 = \alpha) \quad . \tag{3.7}$$

Then Eq. (3.6) defines a recursion

$$\pi_{\alpha n} = g_\alpha(\pi_{s,n-1}, \pi_{m,n-1}) \quad , \quad n \geq 1 \, , \tag{3.8}$$

with

$$\pi_{s0} = \pi_{m0} = 0 \quad .$$

a)   When $\alpha = M$, we have from Eqs. (3.4a) and (3.8)

$$\pi_{mn} = (p_m \pi_{m,n-1} + 1 - p_m)^2 \quad , \quad n \geq 1 \tag{3.9}$$

$$= [p_m(p_m \pi_{m,n-2} + 1 - p_m)^2 + 1 - p_m]^2 \quad , \quad n \geq 2$$

$$= \cdots \quad .$$

The recursion, Eq. (3.9), has the following interpretation. To obtain extinction by generation n, an offspring of a macrophage progenitor either dies, with probability $1 - p_m$, or remains a macrophage progenitor, with probability p_m. In the later case, extinction must occur in the remaining $n - 1$ generations.

With successive iterations, the expression for π_{mn} becomes unwieldy, and so we do not attempt to present it in analytical form. However, as shown in Chapter 7, this expression is useful for numerical evaluations of π_{mn}.

b)   When $\alpha = S$, we have from Eqs. (3.4b) and (3.8)

$$\pi_{sn} = [p_s \pi_{s,n-1} + (1 - p_s)\pi_{m,n-1}]^2 \quad , \quad n \geq 1$$

$$= \{p_s[p_s \pi_{s,n-2} + (1 - p_s)\pi_{m,n-2}]^2 + (1 - p_s)[p_m \pi_{m,n-2} + 1 - p_m]^2\}^2 \quad ,$$

$$n \geq 2$$

$$= \cdots \quad , \tag{3.10}$$

where π_{mn} is given by Eq. (3.9). As in the case $\alpha = M$, we do not attempt to present an analytical expression for π_{sn} but will use Eq. (3.10) to obtain numerical results in Chapter 7.

3.2.2 Eventual completion of growth

The limits as $n \to \infty$ of π_{sn} and π_{mn} are of considerable interest since they measure the likelihood that a cell line, beginning with a stem cell or macrophage progenitor (respectively), will eventually die out.

Define

$$\lim_{n \to \infty} \pi_{sn} = \pi_s \quad \text{and} \quad \lim_{n \to \infty} \pi_{mn} = \pi_m \quad .$$

To see that the limit π_m exists, we use Eq. (3.9) to obtain

$$\pi_{m,n+1} - \pi_{mn} = p_m^2(\pi_{mn}^2 - \pi_{m,n-1}^2) + 2p_m(1 - p_m)(\pi_{mn} - \pi_{m,n-1}) \quad .$$

We can easily establish by induction that π_{mn} is a monotonically increasing sequence. Further, because π_{mn} are probabilities calculated by iterating probability generating functions, $\pi_{mn} \leq 1$. The monotonically increasing sequence π_{mn} is bounded above and thus must converge to a limit. Similarly, from Eq. (3.10)

$$\pi_{s,n+1} - \pi_{sn} = p_s^2(\pi_{sn}^2 - \pi_{s,n-1}^2) + 2p_s(1 - p_s)(\pi_{sn}\pi_{mn} - \pi_{s,n-1}\pi_{m,n-1})$$

$$+ (1 - p_s)^2(\pi_{mn}^2 - \pi_{m,n-1}^2) \quad .$$

Having established that $\pi_{mn} \geq \pi_{m,n-1}$, another inductive argument shows that $\pi_{s,n+1} \geq \pi_{sn}$. As with π_{mn}, the sequence π_{sn} is bounded above by one. Hence it also must converge to a limit. The limits π_s and π_m therefore exist, and

$$\pi_s \leq 1 \quad \text{and} \quad \pi_m \leq 1 \quad .$$

In the remainder of this section, we determine necessary and sufficient conditions for $\pi_s = \pi_m = 1$. When these conditions do not hold, we find explicit expressions for π_s and π_m in terms of p_s and p_m.

The values of the limits π_s and π_m depend on the means of the offspring distribution for S and M parents. We define a matrix, $\mathbf{M}$, whose (i,j)th entry gives the mean number of offspring of type j contributed by a parent of type i in one generation. Thus, for our modified process,

$$\mathbf{M} = \begin{array}{c} \\ S \\ M \end{array} \begin{pmatrix} \overset{S}{2p_s} & \overset{M}{2(1-p_s)} \\ 0 & 2p_m \end{pmatrix}. \tag{3.11}$$

A branching process with a matrix of means having this form is known as a *reducible process*. (In contrast, an *irreducible* branching process is one in which every type of individual can give rise either directly or indirectly to every other type of individual.) The process is also *non-singular*. (A *singular* multitype branching process is one in which every type of individual has exactly one child, regardless of type of child.) [The reader is referred to Harris (1963) and Mode (1971) for a further discussion of classification of processes and matrices of means.] By Theorem 10.1 of Harris (1963) we obtain our first desired result:

THEOREM 1

(a) For a colony having a stem cell parent, completion of growth is certain if, and only if, the probability of self-replication of both S and M cells is at most 0.5, i.e.,

$$\pi_s = 1 \quad <=> \quad \max(p_s, p_m) \le 0.5 \ .$$

(b) For a colony having a macrophage progenitor parent, completion of growth is certain if, and only if, the probability of self-replication of M cells is at most 0.5, i.e.,

$$\pi_m = 1 \quad <=> \quad p_m \le 0.5 \ .$$

PROOF

(a) To apply Theorem 10.1 of Harris, we require the maximum eigenvalue, ρ, of the matrix M. For M, given by Eq. (3.11), the eigenvalues are $\rho_1 = 2p_s$ and $\rho_2 = 2p_m$. Then, Theorem 10.1 states that for a non-singular reducible multitype branching process,

$$\rho = \max(\rho_1, \rho_2) \leq 1 \quad \Longleftrightarrow \quad \pi_s = 1 \quad .$$

But $\rho \leq 1 \Longleftrightarrow \max(p_s, p_m) \leq 0.5$, thus proving (a).

(b) If we modify the two-type (M, E) process to allow offspring of an M-cell parent to be of type M or to die, then the modified process is a one-type process with branching probability p_m. A well-known result (cf. Karlin and Taylor, 1975) states that in a single-type process, extinction is certain if, and only if, $p_m \leq 0.5$. Thus (b) is proven.

For colonies initiated by a stem cell, the magnitude of ρ, the maximum eigenvalue of the matrix M, determines into which of three general growth rate classes the process falls. If ρ is less than 1, then the process is described as *subcritical*, for the growth rate is below the level required to have positive probability of sustaining unending growth. If ρ is greater than 1, then the process is described as *supercritical*, for a colony now has a positive probability of growing without bound. The case $\rho = 1$ is described as *critical*. It provides a threshold at which the characteristics of colony growth change qualitatively and quantitatively. Colonies initiated by M cells may be classified in the same manner, with $\rho = 2p_m$.

For supercritical processes, the extinction probability $\pi_\alpha < 1$. The precise value of π_α is a function of the branching probabilities. To calculate π_α, consider first a cell line initiated by an M cell. From Eq. (3.9), we obtain

$$\pi_m = \lim_{n \to \infty} \left[p_m^2 \pi_{m,n-1}^2 + 2p_m(1 - p_m)\pi_{m,n-1} + (1 - p_m)^2 \right]$$

$$= p_m^2 \pi_m^2 + 2p_m(1 - p_m)\pi_m + (1 - p_m)^2 \quad . \tag{3.12}$$

Comparing with Eq. (3.4a), we see that π_m is a root of

$$\pi_m = g_m(\pi_m) \quad . \tag{3.13}$$

Two roots of Eq. (3.13) exist. By inspection of Eq. (3.12), $\pi_m = 1$ is one root. By Theorem 1 this solution obtains if, and only if, $p_m \le 0.5$. When $p_m > 0.5$, we solve the quadratic equation (3.12) to obtain the second root,

$$\pi_m = \left(\frac{1 - p_m}{p_m}\right)^2 \quad , \quad 0.5 \le p_m \le 1 \quad . \tag{3.14}$$

We use Eq. (3.10) and apply the same approach to find π_s. It can be proven that as $n \to \infty$, the product $\pi_{sn}\pi_{mn} \to \pi_s\pi_m$ (cf. Thomas, 1963). Hence, from Eq. (3.10),

$$\pi_s = \lim_{n \to \infty} \left[p_s^2\pi_{s,n-1}^2 + 2p_s(1 - p_s)\pi_{s,n-1}\pi_{m,n-1} + (1 - p_s)^2\pi_{m,n-1}^2 \right]$$

$$= p_s^2\pi_s^2 + 2p_s(1 - p_s)\pi_m\pi_s + (1 - p_s)^2\pi_m^2 \quad . \tag{3.15}$$

Since π_m is known independently as a function of p_m, Eq. (3.15) becomes a quadratic equation in π_s, with one physically realizable solution

$$\pi_s = \frac{1}{2p_s^2} - \left(\frac{1 - p_s}{p_s}\right)\pi_m - \frac{1}{p_s}\sqrt{\frac{1}{4p_s^2} - \left(\frac{1 - p_s}{p_s}\right)\pi_m} \quad . \tag{3.16}$$

(The solution using the positive radical is greater than 1 for $0 < \pi_m < 1$, and hence can be rejected.) When $\pi_m = 1$, Eq. (3.15) is formally equivalent to Eq. (3.12) and hence $\pi_s = [(1 - p_s)/p_s]^2$.

To summarize our results:

THEOREM 2

Let π_α be the probability that a colony with initial parent of type α ($\alpha = S, M$) reaches completion. Let p_α be the probability that an offspring of a parent of type α is itself of type α. Then

$$\pi_m = \begin{cases} 1 , & 0 < p_m \leq 0.5 \\ \left(\dfrac{1 - p_m}{p_m}\right)^2, & 0.5 \leq p_m \leq 1 \end{cases} \tag{3.17a}$$

and

$$\pi_s = \begin{cases} 1 , & 0 < p_s \leq 0.5 , \quad 0 < p_m \leq 0.5 \\ \left(\dfrac{1 - p_s}{p_s}\right)^2, & 0.5 \leq p_s \leq 1 , \quad 0 < p_m \leq 0.5 \\ \dfrac{1}{2p_s^2} - \left(\dfrac{1 - p_s}{p_s}\right)\left(\dfrac{1 - p_m}{p_m}\right)^2 \\ \qquad - \dfrac{1}{p_s}\sqrt{\dfrac{1}{4p_s^2} - \left(\dfrac{1 - p_s}{p_s}\right)\left(\dfrac{1 - p_m}{p_m}\right)^2} , & 0 < p_s \leq 1 , \quad 0.5 \leq p_m \leq 1 , \end{cases} \tag{3.17b}$$

where we have substituted for π_m from Eq. (3.14) into Eq. (3.16) to obtain the last equality.

3.3   Stem Cell Growth

The fate of the stem cell compartment over time deserves particular attention, since these cells are vital for continued replenishment of the hemopoietic system *in vivo*. We shall characterize the stem cell compartment by finding the probability distribution of its size in finite time and then asking for the probability that a colony initiated by a stem cell still contains a stem cell at later generations.

Our approach is to design a one-type process: We assume stem cells produce offspring of their own type or offspring that "die," i.e., differentiate. The offspring distribution has probability generating function

$$h_s(\theta) = (p_s\theta + 1 - p_s)^2 , \quad 0 < |\theta| \leq 1 . \tag{3.18}$$

3.3.1 Finite time

If we iterate Eq. (3.18) a finite number of times, say n, the coefficient of θ^k in the resulting polynomial of degree $2n$ will give the probability that there are k stem cells in a colony after n generations. For example, the second iterate of Eq. (3.18) is

$$
\begin{aligned}
h_s^{(2)}(\theta) &= [p_s h_s(\theta) + 1 - p_s]^2 \\
&= [p_s(p_s\theta + 1 - p_s)^2 + 1 - p_s]^2 \\
&= p_s^6\theta^4 + 4p_s^5(1 - p_s)\theta^3 + [6p_s^4(1 - p_s)^2 + 2p_s^2(1 - p_s)]\theta^2 \\
&\quad + [4p_s^3(1 - p_s)^3 + 4p_s^2(1 - p_s)^2]\theta + [p_s(1 - p_s)^2 + (1 - p_s)]^2
\end{aligned}
$$

from which we obtain the complete probability distribution of $[Z_{s2}|Z_0]$ as

$$
Pr[Z_{s2} = k|Z_0 = S] = \begin{cases}
p_s^6, & k = 4 \\
4p_s^5(1 - p_s), & k = 3 \\
6p_s^4(1 - p_s)^2 + 2p_s^3(1 - p_s), & k = 2 \\
4p_s^3(1 - p_s)^3 + 4p_s^2(1 - p_s)^2, & k = 1 \\
[p_s(1 - p_s)^2 + (1 - p_s)]^2, & k = 0 \ .
\end{cases}
$$

Further iterations can easily be done with a symbolic manipulation computer program such as MACSYMA (Symbolics, Inc.).

As a direct consequence of obtaining the probability of $[Z_{sn}|Z_0 = S]$, we can obtain the probability that the stem cell compartment contains at least one stem cell and therefore, in principle, retains the capacity for self-renewal. To proceed we note that

$$
Pr[Z_{sn} \geq 1|Z_0 = S] = 1 - Pr[Z_{sn} = 0|Z_0 = S] \ .
$$

Further, observe that Eq. (3.18) has the same functional form as the generating function equation, Eq. (3.4a), for macrophages. Thus, we recognize that our treatment of the stem cell compartment is equivalent to the modified process introduced in Section 3.2 to examine the probability of completion of growth of colonies initiated by a macrophage progenitor. Hence, $Pr[Z_{sn} = 0|Z_0 = S]$ is equivalent to π_{mn}, given by Eq. (3.9), provided we substitute p_s for p_m.

3.3.2 Infinite time

Eventually, limitations of computing capacity will prevent further iteration of $h_s(\theta)$. However, we can still obtain information about the likelihood of the continued presence of stem cells in the colony by calculating the probability of extinction in a one-type process. The probability of extinction of stem cells in this process is found in exactly the same way as the probability of completion of growth for a colony beginning with a macrophage progenitor. Hence, from Eq. (3.17a) with p_s substituted for p_m, we obtain

$$\lim_{n \to \infty} Pr[Z_{sn} = 0|Z_0 = S] = \begin{cases} 1 \, , & 0 < p_s \leq 0.5 \\ \left(\dfrac{1 - p_s}{p_s}\right)^2 \, , & 0.5 \leq p_s \leq 1 \, . \end{cases} \tag{3.19}$$

The long-term probability that a colony will contain at least one stem cell is therefore $\max\left(0, 1 - [(1 - p_s)/p_s]^2\right) = \max(0, \ (2p_s - 1)/p_s^2)$.

Chapter 4

Colonies Reaching Completion

Two related quantities that describe a colony that completes its growth are the time (number of generations) taken to reach completion and the colony size at completion. One might expect that a colony that grows for many generations before its growth finally ceases would be much larger than a colony that ceases growth after only a few generations. We examine first the distribution of the number of generations to completion since the derivation of the distribution follows directly from the finite-time recursion of the modified model in Section 3.2. Colony size at completion is covered later.

4.1   Number of Generations to Completion

In the two-type process of Section 3.2 in which macrophage produce offspring that are either of their own type or that die, we defined

$$Y_{\alpha n} = \text{number of cells of type } \alpha \text{ in the colony at time } n \geq 0 \ , \quad \alpha = S, M \ ,$$

and

$$\mathbf{Y}_n = (Y_{sn}, Y_{mn}) \quad .$$

Now, also define

$$N_\alpha = \text{number of generations to completion of colony growth, given } \mathbf{Y}_0 = \alpha \ .$$

Then

$$Pr(N_\alpha \leq c) = Pr(\mathbf{Y}_c = 0 | \mathbf{Y}_0 = \alpha) \quad .$$

Hence

$$Pr(N_\alpha = c) = Pr(\mathbf{Y}_c = 0 | \mathbf{Y}_0 = \alpha) - Pr(\mathbf{Y}_{c-1} = 0 | \mathbf{Y}_0 = \alpha) \ , \tag{4.1}$$

which, by Eq. (3.7), becomes

$$Pr(N_\alpha = c) = \pi_{\alpha c} - \pi_{\alpha,c-1} \quad . \tag{4.2}$$

Notice that $\lim_{c \to \infty} Pr(N_\alpha \leq c) = 1$ if, and only if, $\pi_\alpha = 1$.

Using Eq. (4.2) we now find when $\alpha = M$, Eq. (3.9) gives the probability that the number of generations to completion of growth is c as

$$Pr(N_m = c) = p_m^2 \pi_{m,c-1}^2 + [2p_m(1 - p_m) - 1]\,\pi_{m,c-1} + (1 - p_m)^2 \quad ; \tag{4.3}$$

whereas, when $\alpha = S$, Eq. (3.10) gives

$$Pr(N_s = c) = p_s^2 \pi_{s,c-1}^2 + [2p_s(1 - p_s)\pi_{m,c-1} - 1]\pi_{s,c-1} + (1 - p_s)^2 \pi_{m,c-1}^2 \quad . \tag{4.4}$$

4.2 Colony Size at Completion

To obtain more detail in the characterization of colony growth, we examine the size of the colonies at completion. When growth is complete, the only cell type present is the end cell. Hence, in the notation of Section 2.1, we seek

$$Pr(Z_n = (0,0,k)|Z_0 = \alpha) \quad , \quad \alpha = S, M \quad ; \quad k = 1, 2, \ldots \quad ; \quad n = 1, 2, \ldots \quad .$$

Notice that $Z_n = (0, 0, k)$ does not imply that the colony ceases growing at the nth generation. Rather, the colony ceases growing at or before the nth generation, at which time it contains k end cells. By finding the minimum value of n at which there are no stem cells or macrophages present one can determine the generation at which the colony ceases growing.

4.2.1 Finite-Time

To obtain $Pr(Z_n = (0,0,k)|Z_0 = \alpha)$ for $n = 1, 2, \ldots$, and different values of k and α, we must iterate the vector–valued generating function, $\mathbf{f}(\boldsymbol{\theta})$, n times and then extract coefficients from the resulting polynomial in θ. This is a daunting exercise to carry out analytically or even computationally. In the nth generation ($n \geq 2$), k can take on up to 2^n possible values. For each value of k, the coefficient of $\theta_s^0 \theta_m^0 \theta_e^k$ is a polynomial in p_s and p_m. Thus, as n increases, the generating functions $\mathbf{f}^{(n)}(\boldsymbol{\theta})$ rapidly increase in complexity. By choosing particular values for p_s and p_m, the coefficient of $\theta_s^0 \theta_m^0 \theta_e^k$ is a number, and the generating function can then be iterated for small n using a symbolic manipulation computer program such as MACSYMA (Symbolics, Inc.) (see Chapter 7). For comparison with many experimental results and in parameter estimation problems, such numerical results are of use.

4.2.2 Infinite-Time

Although determining the final colony size at some finite time n is difficult, it is rather straightforward to determine the final colony size in the limit as $n \to \infty$. Because cells do not die, examining this limit will tell us the ultimate colony size but will not provide information as to when this final size was reached. To proceed, let $a_\alpha(k)$ be the probability that a colony with a parent of type α ultimately forms a colony composed only of k end cells. Thus

$$a_\alpha(k) = \lim_{n \to \infty} Pr(\mathbf{Z}_n = (0, 0, k) | \mathbf{Z}_0 = \alpha) \quad , \quad \alpha = S, M, E \quad . \tag{4.5}$$

Then

$$\psi_\alpha(\theta) = \sum_{k=0}^{\infty} a_\alpha(k)\theta^k \quad , \quad \alpha = S, M, E \quad , \quad |\theta| \leq 1 \tag{4.6}$$

is the probability generating function for the number of end cells in a colony with a parent of type α that has reached completion. Note, in particular, that

$$\psi_e(\theta) = \theta \quad . \tag{4.7}$$

We use an adaptation of Theorem 10.2 of Harris (1963), given below, to calculate $a_\alpha(k)$. In order to use this theorem, we need to show that E is a final group of the process. The reader may check this rigorously by referring to Harris. However, the essence of a final group is that it acts as a "sink." E clearly fulfills this function in our model. We must also recall the definitions of the probability generating functions from Section 2.1, namely

$$f_s(\theta) = [p_s\theta_s + (1 - p_s)\theta_m]^2$$

and

$$f_m(\theta) = [p_m\theta_m + (1 - p_m)\theta_e]^2 \quad .$$

THEOREM 3 (Harris, 1963, Theorem 10.2)

Suppose $f_\alpha(0) = 0$, $\alpha = S, M$. Then the functions $\psi_\alpha(\theta)$ are uniquely determined by the equations

$$\psi_s(\theta) = f_s(\psi_s(\theta), \psi_m(\theta))$$

$$(4.8)$$

$$= [p_s\psi_s(\theta) + (1 - p_s)\psi_m(\theta)]^2$$

and

$$\psi_m(\theta) = f_m(\psi_m(\theta), \psi_e(\theta))$$

$$(4.9)$$

$$= [p_m\psi_m(\theta) + (1 - p_m)\psi_e(\theta)]^2 \quad .$$

To use this theorem to solve for $a_\alpha(k)$, we express both sides of Eqs. (4.8) and (4.9) as polynomials in θ and equate coefficients of θ^k. We begin by expanding Eq. (4.9) using Eqs. (4.6) and (4.7), to obtain

$$\sum_{k=0}^{\infty} a_m(k)\theta^k = p_m^2 \left[\sum_{k=0}^{\infty} a_m(k)\theta^k\right]^2 + 2p_m(1 - p_m)\left[\sum_{k=0}^{\infty} a_m(k)\theta^{k+1}\right] + (1 - p_m)^2\theta^2$$

$$= p_m^2 \left[\sum_{k=0}^{\infty}\sum_{j=0}^{k} a_m(j)a_m(k - j)\theta^k\right]$$

$$+ 2p_m(1 - p_m)\left[\sum_{k=1}^{\infty} a_m(k - 1)\theta^k\right] + (1 - p_m)^2\theta^2 \quad . \tag{4.10}$$

Now,

$$a_m(0) = \lim_{n \to \infty} Pr(Z_n = 0 | Z_0 = M) = 0 \quad ,$$

since at least two end cells must be produced from a single macrophage. For the same reason, $a_m(1) = 0$. Equating coefficients of θ^k gives the following recursion:

$$a_m(k) = p_m^2 \left[\sum_{j=0}^{k} a_m(j)a_m(k-j) \right] + 2p_m(1-p_m)a_m(k-1) + (1-p_m)^2 \delta_{k,2} \ , \quad k \geq 2 \ , \quad (4.11a)$$

with

$$a_m(0) = a_m(1) = 0 \ . \tag{4.11b}$$

By comparing with Eq. (3.1b), we see that Eq. (4.11) is simply the limiting form of Eq. (3.1b) as $n \to \infty$. The reason for the equivalence is that, after an infinite number of generations, those colonies that will cease growing have done so. Hence, the total colony size, irrespective of cell type, is equivalent to the total number of end cells. However, for any n, Eq. (3.1b) defines a non-defective probability distribution, while Eq. (4.11) above defines a defective probability distribution under conditions in which the probability of completion of growth is less than one (see Section 3.2). Formally, we have

$$\sum_{k=0}^{\infty} a_m(k) = \pi_m \leq 1 \ . \tag{4.12}$$

Calculations for $a_s(k)$ proceed in a similar fashion. Again, the resulting expression is simply the limit as $n \to \infty$ of Eq. (3.1a), namely

$$a_s(k) = p_s^2 \left[\sum_{j=0}^{k} a_s(j)a_s(k-j) \right] + 2p_s(1-p_s) \left[\sum_{j=0}^{k} a_s(j)a_m(k-j) \right]$$

$$\tag{4.13a}$$

$$+ (1-p_s)^2 \left[\sum_{j=0}^{k} a_m(j)a_m(k-j) \right] \ , \quad k \geq 2 \ ,$$

with

$$a_s(0) = a_s(1) = 0 \tag{4.13b}$$

and $a_m(k)$, $k \geq 0$, given by Eq. (4.11). Again

$$\sum_{k=0}^{\infty} a_s(k) = \pi_s \leq 1 \ . \tag{4.14}$$

The probabilities $a_\alpha(k)$ describing the colony size at completion, in conjunction with the probabilities $Pr(N_\alpha \leq c)$ describing the number of generations to completion, provide information about colonies that cease growing. Together, these two distributions lend insight into the process that is described completely by the distribution $Pr(\mathbf{Z}_n = (0, 0, k)|\mathbf{Z}_0)$, which we have not been able to compute in general. Unfortunately, it is only through this last distribution that we can carefully examine the correlation between growing time and colony size at completion of growth.

Chapter 5

Colonies Growing without Bound

From Theorem 2 we can predict that colonies containing cell types with branching probabilities greater than 0.5 (i.e., colonies having supercritical growth) will, with non-zero probability, never reach completion of growth. When growth is unbounded, calculating the number
of cells in each of the three compartments as $n \to \infty$ is a meaningless exercise, for these numbers will be infinite or zero. (For example, in a colony with an S-cell parent, if $\frac{1}{2} < p_s < 1$
and $p_m < 1$, then there is a non-zero probability that the S, M and E compartments will
each contain an infinite number of cells.) However, we can calculate the proportion of cells
in each of the compartments by appealing to the Kesten-Stigum limit theorems for decomposable Galton-Watson processes (cf. Kesten and Stigum, 1967; Mode, 1971). The essence
of these theorems is that, under conditions in which a colony can grow without bound, the
limiting proportion of each cell type is a constant that can be calculated as a function of the
branching probabilities, p_s, p_m, and p_e. The theorems require calculation of the eigenvalues
and eigenvectors of the matrix of means of the offspring distribution for parents of each type.
We shall consider separately the cases of colonies with an M-cell parent and colonies with an
S-cell parent. Since the former case is easier and is in some sense a subset of the latter case,
we shall consider it first.

5.1 Asymptotic Proportions of Cell Types in Colonies with an M-Cell Parent

In order for a colony with an M-cell parent to have a positive probability of growing
without bound, p_m must exceed 0.5. If we define a matrix of the means of the offspring
distributions, **M**, with elements

$$M_{ij} = \text{mean number of offspring of type } j \text{ contributed by a parent of type } i$$
$$\text{in one generation} \quad , \quad i, j = M, E \ ,$$

then

$$\mathbf{M} = \begin{array}{c} \\ M \\ \\ E \end{array} \begin{array}{c} \overset{M \qquad\qquad E}{\left(\begin{array}{cc} 2p_m & 2(1-p_m) \\ \\ 0 & 1 \end{array} \right)} \end{array} . \qquad (5.1)$$

The limiting behavior of this two-type process depends on the maximum eigenvalue, ρ, of $\mathbf{M}$ and the left and right eigenvectors corresponding to ρ. The eigenvalues of $\mathbf{M}$ are $2p_m$ and 1. Because we are restricting our attention to cases in which $p_m > 0.5$, the maximum eigenvalue, $\rho = 2p_m > 1$.

The left eigenvector corresponding to ρ is

$$\mathbf{v} = \left(\frac{2p_m - 1}{2(1 - p_m)} \;,\; 1 \right) v \;,\quad v \text{ an arbitrary constant} > 0 ,$$

and the right eigenvector corresponding to ρ is $\mathbf{u}'$, where

$$\mathbf{u} = (1,0)u \;,\quad u \text{ an arbitrary constant} > 0 ,$$

with $\mathbf{u}'$ being the transpose of $\mathbf{u}$. We choose the constants u and v so that $\mathbf{vu}' = 1$. Then, Theorem 2.1 of Kesten and Stigum (1967) takes the following form:

THEOREM 4

For $\mathbf{Z}_0 = M$ and $p_m > 0.5$, there exists a random variable W such that

$$\lim_{n \to \infty} \frac{\mathbf{Z}_n}{(2p_m)^n} = W\mathbf{v} \;,$$

where $\mathbf{v} = ((2p_m - 1)/[2(1 - p_m)], 1)v$. Also,

$$E(W|\mathbf{Z}_0 = M) = u > 0 \;,$$

where $uv(2p_m - 1)/[2(1 - p_m)] = 1$ if, and only if,

$$E(Z_{m1} \; log \; Z_{m1}|\mathbf{Z}_0 = M) < \infty \quad . \qquad (5.2)$$

Finally, if Eq. (5.2) holds and if $Z_{m1}u$ can take at least two values with positive probability, given $Z_0 = M$, then the distribution of W has a jump of magnitude $\pi_m = [(1 - p_m)/p_m]^2$ at the origin (i.e., $Pr(W = 0) = \pi_m$), and a continuous density function on the set of positive real numbers.

One can readily check that condition (5.2) is met from the distribution of $Z_{m1}|Z_0 = M$, namely

$$(Z_{m1}|Z_0 = M) = \begin{cases} 0 & \text{with probability} & (1 - p_m)^2 \\ 1 & \text{with probability} & 2p_m(1 - p_m) \\ 2 & \text{with probability} & p_m^2 \end{cases} \qquad (5.3)$$

where Z_{m1} is the number of macrophages in the first generation. It is also true from this distribution that $Z_{m1}u$ can take on at least two values with positive probability provided $p_m \neq 0, 1$.

The random variable, W, represents the stochastic nature of colony growth until the asymptotic behavior is attained. At low generation number, fluctuations in the actual numbers of offspring of individual parents are substantial, relative to the total colony size. However, after many generations, the behavior of the colony as a whole is the sum of behaviors of many individual parents. Now, individual fluctuations compensate for each other so that the overall behavior of the process approaches a steady state with increasing numbers of generations. Some explanation of the mixed discrete and continuous nature of the distribution of W is in order. Consider the distribution of Z_n. As $n \to \infty$, some of the colonies beginning with a macrophage parent will continue to grow without bound, at a per-generation rate of roughly on order $2p_m$. Hence, when those colonies that continue to grow without bound are normalized by $(2p_m)^n$, we obtain $\lim_{n \to \infty} \frac{Z_n}{(2p_m)^n} > 0$. On the other hand, with probability π_m, a colony will reach completion of growth in a finite time. For those colonies reaching completion, $\lim_{n \to \infty} \frac{Z_n}{(2p_m)^n} = 0$ since $2p_m > 1$, and Z_n is bounded once completion is reached. Hence, the jump at the origin in the distribution of W represents the contribution to the normalized limiting colony size of all those colonies eventually reaching completion. Unfortunately, the density function of W is not obtainable analytically for a multitype process.

From Theorem 4, we infer that either, with probability $\pi_m = [(1-p_m)/p_m]^2$, a colony with a macrophage parent reaches completion or, with probability $(1-\pi_m)$, the colony grows without bound, with macrophages and end cells occurring in the constant ratio of $\dfrac{(2p_m-1)}{2(1-p_m)}$: 1. One can show algebraically that

$$\frac{2p_m-1}{2(1-p_m)} > 1 \quad <=> \quad 0.75 < p_m \le 1 \ . \tag{5.4}$$

Consequently, if the colony continues growing, the majority of cells will be end cells except in the situation where macrophages have a probability of self-replacement $p_m > 0.75$, which, in our experience, is uncommonly high.

5.2 Asymptotic Proportions of Cell Types in Colonies with an S-Cell Parent

Colonies arising from an S-cell parent will have positive probability of growing without bound if either or both of p_s and p_m exceeds 0.5. The limiting behavior is described in a theorem, which is the extension of Theorem 4 above, and is again derived from Kesten and Stigum, Theorem 2.1.

As for Theorem 4, Theorem 5 below depends on the maximum eigenvalue with its corresponding left and right eigenvector for the matrix of means of the offspring distributions, **M**. Using the definition of M_{ij} from the previous section, we obtain

$$\mathbf{M} = \begin{array}{c} \\ S \\ M \\ E \end{array} \begin{array}{ccc} S & M & E \\ \left(\begin{array}{ccc} 2p_s & 2(1-p_s) & 0 \\ 0 & 2p_m & 2(1-p_m) \\ 0 & 0 & 1 \end{array} \right) \end{array} \ . \tag{5.5}$$

The results of Kesten and Stigum Theorem 2.1 strictly apply to 2×2 triangular matrices such as that of Eq. (5.1). However, as long as the maximum eigenvalue of the matrix is unique, Theorem 2.1 with easy reinterpretations is true, regardless of the dimensionality of the matrix (Kesten and Stigum, 1967). The following theorem is the extension of Theorem 4 to the case of $\mathbf{M}$, a 3×3 triangular matrix, where $\mathbf{M}$ is given by Eq. (5.5). Let ρ be the maximum eigenvalue of $\mathbf{M}$ with corresponding left and right eigenvectors $\mathbf{v}$ and $\mathbf{u}'$, respectively, with $\mathbf{vu}' = 1$. Following the statement of the Theorem, particular values of ρ, $\mathbf{v}$, and $\mathbf{u}$ are given in two cases: (i) $p_s > 0.5$, $p_s > p_m$ and (ii) $p_m > 0.5$, $p_m > p_s$. The case $p_s = p_m > 0.5$ is not treated because it is highly unlikely to occur in practice, and an unwarranted amount of mathematical notation must be introduced in order to utilize Kesten and Stigum's theory.

THEOREM 5

For $\mathbf{Z}_0 = S$ and $\rho = max\ (2p_s, 2p_m) > 1$, there exists a random variable W such that,

$$\lim_{n \to \infty} \frac{\mathbf{Z}_n}{\rho^n} = W\mathbf{v}\quad .$$

Also,

$$E(W|\mathbf{Z}_0 = S) = u_s > 0\ ,$$

where u_s is the first element of $\mathbf{u}$ if, and only if,

$$E(Z_{s1}\ log\ Z_{s1}|\mathbf{Z}_0 = S) < \infty\ ;\quad\quad\quad (5.6a)$$

$$E(Z_{m1}\ log\ Z_{m1}|\mathbf{Z}_0 = M) < \infty\ .\quad\quad\quad (5.6b)$$

Finally, if Eq. (5.6) holds, and if $u_s(Z_{s1}|\mathbf{Z}_0 = S)$ and $u_m(Z_{m1}|\mathbf{Z}_0 = M)$, where u_m is the second element of $\mathbf{u}$, each can take at least two values with positive probability, then the distribution of W has a jump of magnitude π_s, given by Eq. (3.17b), at the origin and a continuous density function on the set of positive real numbers.

One can readily check that all conditions of the theorem are met by using the distributions

$$Pr(Z_{s1} = i|\mathbf{Z}_0 = S) = \binom{2}{i}p_s^i(1 - p_s)^{2-i}\ ,\quad i = 0, 1, 2\quad\quad (5.7a)$$

and

$$Pr(Z_{m1} = i|\mathbf{Z}_0 = M) = \binom{2}{i}p_m^i(1 - p_m)^{2-i}\ ,\quad i = 0, 1, 2\quad .\quad\quad (5.7b)$$

It remains to calculate ρ, $\mathbf{v}$, and $\mathbf{u}$ in the two cases of interest:

(i) $p_s > p_m$, $p_s > 0.5$

From M in Eq. (5.5) we calculate the following quantities:

$$\rho = 2p_s \ ,$$

$$\mathbf{v} = \left[\left(\frac{p_s - p_m}{1 - p_s} \right) \left(\frac{2p_s - 1}{2(1 - p_m)} \right) \ , \quad \frac{2p_s - 1}{2(1 - p_m)} \ , \quad 1 \right] v \ ,$$

$$v \text{ an arbitrary constant} > 0 \ ,$$

and

$$\mathbf{u} = (1, 0, 0)u \ , \qquad\qquad u \text{ an arbitrary constant} > 0 \ . \qquad (5.8)$$

Hence, $u_s = u$. We choose u and v so that $\mathbf{vu}' = 1$. Thus, the long-term ratio of cells in the S, M, and E compartments is

$$\left(\frac{p_s - p_m}{1 - p_s} \right) \left(\frac{2p_s - 1}{2(1 - p_m)} \right) \ : \ \left(\frac{2p_s - 1}{2(1 - p_m)} \right) \ : \ 1 \ . \qquad (5.9)$$

One can ask for conditions under which the three types achieve particular orderings on the basis of proportions in the colony. If we interpret $\alpha > \beta$ to mean type α occurs in a higher proportion than type β, then

$$S > M \ \text{ if } \ p_s > \frac{1}{2}(1 + p_m) \ ; \qquad (5.10a)$$

$$S > E \ \text{ if } \ p_s > p_m + \frac{1}{4}[\sqrt{16(1 - p_m)^2 + 1} - 1] \ ; \qquad (5.10b)$$

$$M > E \ \text{ if } \ p_s > 1.5 - p_m \ . \qquad (5.10c)$$

The conditions specified in Eq. (5.10) are graphed in Fig. 5.1. These conditions partition the two-dimensional (p_s, p_m) parameter space into regions giving rise to each of the six different orderings of the proportions of each cell type.

(ii) $p_m > p_s$, $p_m > 0.5$

From M in Eq. (5.5) we now calculate

$$\rho = 2p_m \ ,$$

$$\mathbf{v} = \left(0, \frac{2p_m - 1}{2(1 - p_m)}, 1\right) v \ , \quad v \text{ an arbitrary constant } > 0 \ ,$$

and

$$\mathbf{u} = \left(\frac{1 - p_s}{p_m - p_s}, 1, 0\right) u \ , \quad u \text{ an arbitrary constant } > 0 \ .$$

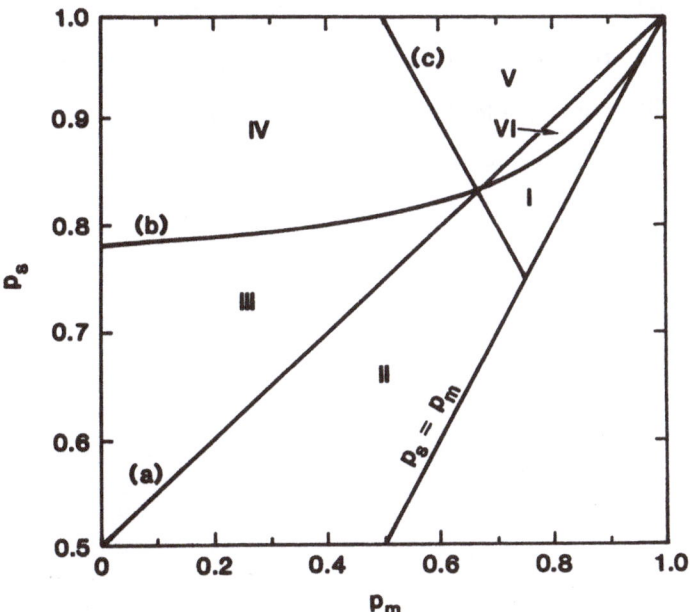

Figure 5.1   Regions of parameter space that give rise to different orderings of proportions of cells of type S, M, and E in supercritical colonies growing from an S-cell parent, $p_s > .5$, $p_s > p_m$. The equations of the lines are given by Eq. (5.10). In the various regions the proportions of cells are as follows: (I) $S < E < M$; (II) $S < M < E$; (III) $M < S < E$; (IV) $M < E < S$; (V) $E < M < S$; (VI) $E < S < M$.

Hence, $u_s = [(1 - p_s)/(p_m - p_s)]u$. We choose u and v so that $\mathbf{vu'} = 1$. Thus, the long-term ratio of cells in the S, M, and E compartments is

$$0 : \frac{2p_m - 1}{2(1 - p_m)} : 1 \quad . \tag{5.11}$$

We note that $p_m > 0.75$ implies that macrophage will always make up greater than half the colony. In contrast, the stem cell proportion approaches zero.

In view of Eq. (5.11), one might ask whether there are stem cells present in the colony. To answer this we can calculate the probability that there is at least one stem cell in the colony in the limit $n \to \infty$. From Eq. (3.19) we find that the long-term probability of containing at least one stem cell in the colony is greater than zero if, and only if, $p_s > 0.5$. Thus, if $p_m > p_s > 0.5$, the proportion of stem cells will approach zero, even though there is a non-zero probability – in fact a probability of $1 - [(1 - p_s)/p_s]^2 = (2p_s - 1)/p_s^2$ – that at least one stem cell is present in the colony.

Chapter 6

Critical Processes

As described in Section 3.2, a threshold exists between colonies completing growth in a finite number of generations and colonies with a positive probability of growing without bound. The condition defining this threshold is that a colony contains cell types for which the maximum branching probability is 0.5 (see Theorem 1). The branching process is then defined as *critical*. We have already seen that a colony with $p_m, p_s \leq 0.5$ has probability 1 of completion of growth. However, if any of the cell types in the colony have a branching probability equal to 0.5, then the expected time for the colony to complete growth is infinite. Thus, aspects of finite and infinite growth are observed in a critical process. One might expect that, biologically, the critical case is important, for it possibly approximates the equilibrium situation *in vivo*.

There is a substantial body of theory related to critical processes. We shall simply introduce a useful result here without delving into the broader aspects of critical processes. The primary case considered below is that in which a colony grows from a stem cell, for which $p_s = 0.5$. Macrophages are also assumed to have branching probability $p_m = 0.5$. (The process with $p_s = p_m = 0.5$ is described as *strongly critical*.) The matrix of means of the offspring distributions, M, defined in Eq. (5.5), thus takes on the form

$$\mathbf{M} = \begin{pmatrix} 1 & 1 & 0 \\ 0 & 1 & 1 \\ 0 & 0 & 1 \end{pmatrix} \quad .$$

We shall also briefly examine the case in which a colony grows from a macrophage with $p_m = 0.5$.

For those colonies initiated by a stem cell, our goal is to describe the long-term growth of the colony, conditional on stem cells being present in the colony. *In vitro*, the stem-cell content of a colony is important, as we shall see in Chapter 7, for it strongly influences the chance of the continued growth of the colony. For those colonies initiated by a macrophage, it is possible to describe the long-term behavior of both macrophages and end cells in a growing colony (i.e., one containing at least one macrophage). The results we use derive from Foster and Ney (1978). Essentially, our approach is the following: For those colonies initiated by a stem cell parent with $p_s = 0.5$, the stem-cell compartment of the process constitutes a critical one-type process. Since the mean time to extinction of a critical one-type process is infinite, the mean time to extinction of the three-type (S, M, and E) process, beginning with a stem cell, is also infinite. A similar argument applies to those colonies initiated by a macrophage with $p_m = 0.5$. From Foster and Ney (1978), we obtain the following:

THEOREM 6

(i) Assume $Z_0 = S$, and $p_s = p_m = 0.5$. Then:

(a)

$$\lim_{n \to \infty} \left\{ \frac{Z_{sn}}{n}, \frac{Z_{mn}}{n^2}, \frac{Z_{en}}{n^3} \middle| Z_{sn} \geq 1 \right\} = (W_s, W_m, W_e) \quad , \tag{6.1}$$

where W_s, W_m, and W_e are proper random variables with $Pr(W_\alpha > 0) > 0$, $\alpha = S$, M, E and

(b)

$$\lim_{n \to \infty} Pr\left(\frac{Z_{sn}}{n} > u \middle| Z_{sn} \geq 1 \right) = e^{-4u} \quad , \quad u > 0 \quad . \tag{6.2}$$

(ii) Assume $Z_0 = M$, and $p_m = 0.5$. Then:

(a)

$$\lim_{n \to \infty} \left\{ \frac{Z_{mn}}{n}, \frac{Z_{en}}{n^2} \middle| Z_{mn} \geq 1 \right\} = (W_m, W_e) \quad , \tag{6.3}$$

where W_m and W_e are proper random variables [different from those of (i)] with $Pr(W_\alpha > 0) > 0$, $\alpha = M$, E and

(b)

$$\lim_{n\to\infty} Pr\left(\frac{Z_{mn}}{n} > u | Z_{mn} \geq 1\right) = e^{-4u} \quad , \quad u > 0 \quad , \tag{6.4}$$

and

$$\lim_{n\to\infty} Pr\left(\frac{Z_{en}}{n^2} > u | Z_{mn} \geq 1\right) = \frac{d\theta_4(0; u)}{du} \quad , \tag{6.5}$$

where $\theta_4(0; u)$ is a theta function (Spanier and Oldham, 1987) defined by

$$\theta_4(0; u) = \frac{1}{\sqrt{\pi u}} \sum_{j=-\infty}^{\infty} exp[-(1/2 + j)^2/u] \quad . \tag{6.6}$$

PROOF

To prove case (i)a, we appeal to Theorem 2, Part (a), of Foster and Ney (1978). We first check that the appropriate conditions are met. To do this we define real constants w_S, w_M, and w_E and the integer-valued vector $\mathbf{k} = (k_S, k_M, k_E)$. We then define the moment

$$\mu(\mathbf{k}) = E\left[\prod_{i=S,M,E} (w_i Z_{i1})^{k_i} | \mathbf{Z}_0 = S\right] \tag{6.7}$$

and must show that the moment exists when $k_S + k_M + k_E \leq 2$. Existence of $\mu(\mathbf{k})$ is obvious since $Z_{i1} \leq 2$ ($i = S, M, E$). The remainder of the proof of Theorem 2(a) of Foster and Ney (1978) depends on the matrix of means of the offspring distribution, $\mathbf{M}$. Foster and Ney (1978) consider a matrix, $\mathbf{M}$, composed of submatrices along and above the diagonal. The diagonal sub-matrices are assumed to have unique maximum eigenvalues, 1, with corresponding left and right eigenvectors that are normalized so that their inner product is 1. In the particular case above, we have three 1×1 matrices along the diagonal, one matrix corresponding to each type in the process. Hence, we can set left and right eigenvectors (of unit length) to 1. In the notation of Foster and Ney's Theorem 2 (we leave details to the reader), $u_i = v_i = 1$, $i = 1, 2, 3$. The result (i)a is now a direct re-expression of Theorem 2(a).

To prove case (i)b one can use Part (b) of Theorem 2, Foster and Ney (1978). This theorem gives a general result for the Laplace transform of the limiting distribution of

(W_s, W_m, W_e) from which (i)b can be derived. However, stating the Foster and Ney result involves the introduction of a substantial amount of notation that is not relevant to the special case that we are considering. In the interests of efficiency we develop a simpler proof of (i)b by appealing to Theorem 10.1, Chapter 1, of Harris (1963) which deals with single type branching processes.

To be specific, consider the stem cell compartment alone. Any non-stem cell offspring are "deaths" from the viewpoint of the stem cell population. Since $p_s = 0.5$, the univariate probability generating function for stem cell production is

$$h_s(\theta_s) = (p_s\theta_s + 1 - p_s)^2 = \frac{1}{4}\theta_s^2 + \frac{1}{2}\theta_s + \frac{1}{4} \quad , \quad |\theta_s| \leq 1 \ . \tag{6.8}$$

Theorem 10.1 of Harris (1963) states the following: Suppose $E[Z_{s1}|Z_0 = S] = 1$ and $h_s'''(1) < \infty$. Then

$$\lim_{n\to\infty} Pr\left(\frac{2Z_{sn}}{nh_s''(1)} > v | Z_{sn} \geq 1\right) = e^{-v} \quad , \quad v \geq 0 \ . \tag{6.9}$$

From Eq. (6.8) it is easy to see that $E[Z_{s1}|Z_0 = S] = 1$ and $h_s'''(1) = 0$. Thus, the conditions of Harris, Theorem 10.1, are satisfied. For our model Eq. (6.9) becomes

$$\lim_{n\to\infty} Pr\left(\frac{Z_{sn}}{n} > \frac{v}{4}|Z_{sn} \geq 1\right) = e^{-v} \quad , \quad v \geq 0 \ . \tag{6.10}$$

Setting $v/4 = u$ gives result (i)b.

To prove case (ii) we observe that macrophages now play the same role as stem cells did in (i). Equations (6.3) and (6.4) thus follow from (i) by a simple change of notation. To prove Eq. (6.5) we note that Foster and Ney (1978) consider a two-type process as an example illustrating their Theorem 2. In their equation (2.16) they give the Laplace transform of the marginal distribution of W_e as

$$L.T.[W_e(u)] = 2\sqrt{\lambda m_{s2}} \operatorname{csch}\left(2\sqrt{\lambda m_{s2}}\right) \ , \tag{6.11}$$

where $L.T.$ denotes Laplace transform, s_2 is the transform variable, and m is the mean number of end cell offspring from a macrophage parent. For the case under consideration, $p_m = 0.5$, and thus $m = 1$. Lastly,

$$\lambda = \frac{1}{2}E[Z_{m1}^2 - Z_{m1}|Z_0 = M] \quad . \tag{6.12}$$

Now

$$(Z_{m1}|Z_0 = M) = \begin{cases} 0 \\ 1 \\ 2 \end{cases} \quad \text{with probability} \quad \begin{cases} \frac{1}{4} \\ \frac{1}{2} \\ \frac{1}{4} \end{cases}$$

since $p_m = 0.5$. Therefore, $\lambda = \frac{1}{4}$. Taking the inverse Laplace transform of Eq. (6.11) (cf. Spanier and Oldham, 1987) with $m = 1$ and $\lambda = 0.25$ yields Eq. (6.5). This completes the proof.

The implication of Theorem 6 is the following: Although strongly critical processes have probability 1 of ceasing growth, by conditioning on the existence of at least one stem cell in the colony, we are restricting attention to colonies that in principle have the capacity for self-renewal. For these colonies, Theorem 6 states that in generation n, for n large, the composition of the colony will be such that the number of stem cells, macrophage progenitors, and end cells, will be proportional to n, n^2, and n^3, respectively, where the "constant of proportionality" for each cell type is a distinct random variable. In the strongly critical case, $Pr(Z_{sn} \geq 1)$ is a monotonically decreasing function of n, which becomes small for large n. For example, $Pr(Z_{sn} \geq 1) = 0.11$ when $n = 30$. [$Pr(Z_{sn} \geq 1)$ is equal to one minus the probability of extinction by generation n for a one-type process. It can be computed from Eq. (3.9) for π_{mn} with $p_m = 0.5$. Values of this function are illustrated in Fig. 7.5.] Thus the chance of finding a colony containing any stem cells approaches zero as $n \to \infty$. However, if a colony does contain a stem cell it will be large, having the number of end cells proportional to n^3 and the number of macrophages proportional to n^2. A similar interpretation can be given to colonies that start with a macrophage parent, except such colonies will be smaller since the number of end cells will only be proportional to n^2.

Chapter 7

Results

We now use the results of our mathematical analyses in the previous chapters to make predictions about the general behavior of colony growth in culture. We will emphasize contrasts between colonies originating from stem cells and colonies originating from more mature cells that lack the capability of self-renewal, which in our model are macrophage progenitors. We will also describe the pattern of growth with increasing time in a heterogeneous culture.

7.1 Parameter Estimation

To illustrate our predictions, we need to choose values for the parameters η_s, η_m, and η_e representing the fraction of cells of types S, M, and E (respectively) that are initially plated, and values for the branching probabilities p_s and p_m representing the probability that a daughter cell is of the same type as its parent for S and M cell parents (respectively; see Fig. 2.1). In Appendix B, we outline a formal procedure for obtaining method of moments estimators of p_s and p_m, conditioned upon preselected values of η_s, η_m, and η_e. To carry out this estimation procedure is a substantial undertaking and one which we choose not to perform in this present study. Our objective here is to explore the general behavior of our model rather than to attempt to fit the model to one or several experimentally generated data sets.

We have chosen values for η_s, η_m, η_e, and p_s, p_m using the following reasoning: Some experimental evidence is available indicating values for η_s, η_m, and η_e. It is generally accepted that stem cells are rare (perhaps 0.1% to 3% of the bone marrow cell population). C. C. Stewart (Los Alamos National Laboratory, personal communication) believes that his experimental technique enriches somewhat for stem cells and estimates that approximately 20% of the cells that he plates never proliferate. Hence, we selected $\eta_s = 0.03$, $\eta_m = 0.77$, $\eta_e = 0.20$.

Direct experimental evidence for realistic values of p_s and p_m is rarer. However, the experimentally observed high frequency of relatively small colonies, containing less than 50 cells at completion of growth, suggests that the predominant parent (i.e., the macrophage progenitor) has a branching probability below the critical value of 0.5 so that the probability that a colony ceases growth is 1 (see Section 3.2). On the other hand, because we know that very large colonies are seen infrequently and are expected to arise from a parent with high proliferative potential, we believe that the case $p_s > 0.5$ is important since, in our model, this case will lead to a positive probability that colonies grow without bound (see Section 3.2). In fact, both Vogel et al. (1968, 1969) and Burgess and Nicola (1983) surmise that the self-renewal probability of a stem cell is close to 0.6.

Hence, many of our predictions are calculated for the values $p_s = 0.6$, $p_m = 0.45$. The choice of $p_m = 0.45$ is to ensure that, although colonies with macrophage progenitor parents exhibit subcritical growth (i.e., have probability 1 of eventually ceasing growth), the difference between colony growth from the two types of parents is not exaggerated. Frequently, it is interesting and informative to examine the results for sensitivity to the choice of p_s and p_m. We therefore also make predictions at or near the critical branching probabilities $p_s = 0.5 = p_m$; i.e., on the boundary between probability 1 and probability less than 1 of ceasing growth. We then compare predictions for qualitative changes in behavior.

In circumstances where we have reason to believe that $p_m \leq 0.5$ and in which we have reasonable estimates for the proportion of initiating cells in each of the three classes, we may use the observed proportion of colonies reaching completion of growth to estimate p_s. The estimation procedure is based on the following:

$$Pr(\text{eventual completion of growth}) = \sum_{\alpha=S,M,E} \eta_\alpha Pr(\text{eventual completion of growth}|Z_0 = \alpha)$$

$$= \eta_s \pi_s + \eta_m + \eta_e \ , \tag{7.1}$$

where π_s is the probability of eventual completion of growth of a colony initiated by a stem cell. Equation (3.17b) gives

$$\pi_s = \left(\frac{1-p_s}{p_s}\right)^2 . \tag{7.2}$$

When $p_m \leq 0.5$, the probability of completion of growth of a colony initiated by a macrophage progenitor $\pi_m = 1$, and thus π_m need not be included in Eq. (7.1).

Let $\hat{P}_c$ be the observed proportion of colonies reaching completion (i.e., that ultimately cease growing). Then p_s is estimated by $\hat{p}_s$, the solution to the following quadratic equation obtained from Eqs. (7.1) and (7.2):

$$\hat{p}_s^2 \left(\frac{\hat{P}_c - \eta_m - \eta_e}{\eta_s} - 1\right) + 2\hat{p}_s - 1 = 0 . \tag{7.3}$$

The only root to this equation that is a valid probability lying between zero and one is

$$\hat{p}_s = \frac{\sqrt{\dfrac{\hat{P}_c - \eta_m - \eta_e}{\eta_s} - 1}}{\dfrac{\hat{P}_c - \eta_m - \eta_e}{\eta_s} - 1} , \tag{7.4}$$

provided that $1 \geq \hat{P}_c > \eta_m + \eta_e$.

7.2 Total Colony Size After n Generations

The most readily obtained experimental data is the colony size distribution, especially for short periods of time when the colonies are small and therefore easy to count. Using Eq. (3.1), we calculated the probability distribution for colony size as a function of time for branching probabilities $p_s = 0.5 = p_m$ and $p_s = 0.6$, $p_m = 0.45$. The results are plotted in Fig. 7.1 for 4, 8, 16, and 32 generations for $p_s = 0.5 = p_m$. Qualitative behavior in the case $p_s = 0.6$, $p_m = .45$ was similar, as shown in Fig. 7.2; the graphs of the probability distribution for colony size as a function of time had the same shape for both sets of branching

Figure 7.1 Probability that colony size $= N$, at generation n for colonies with S-cell, M-cell, or randomly chosen parent; (a) n $=$ 4, (b) n $=$ 8, (c) n $=$ 16, (d) n $=$ 32. Illustrated for $p_s = 0.5 = p_m$. Probability distribution for colonies with a random parent is a weighted average of the distributions for the known parents, where weighting is according to the fraction of cells of each type in the plated cell suspension. The upper tail of the distribution ($N > 50$) is combined into a single category. The mean size of colonies in this category is written beside the probability peak.

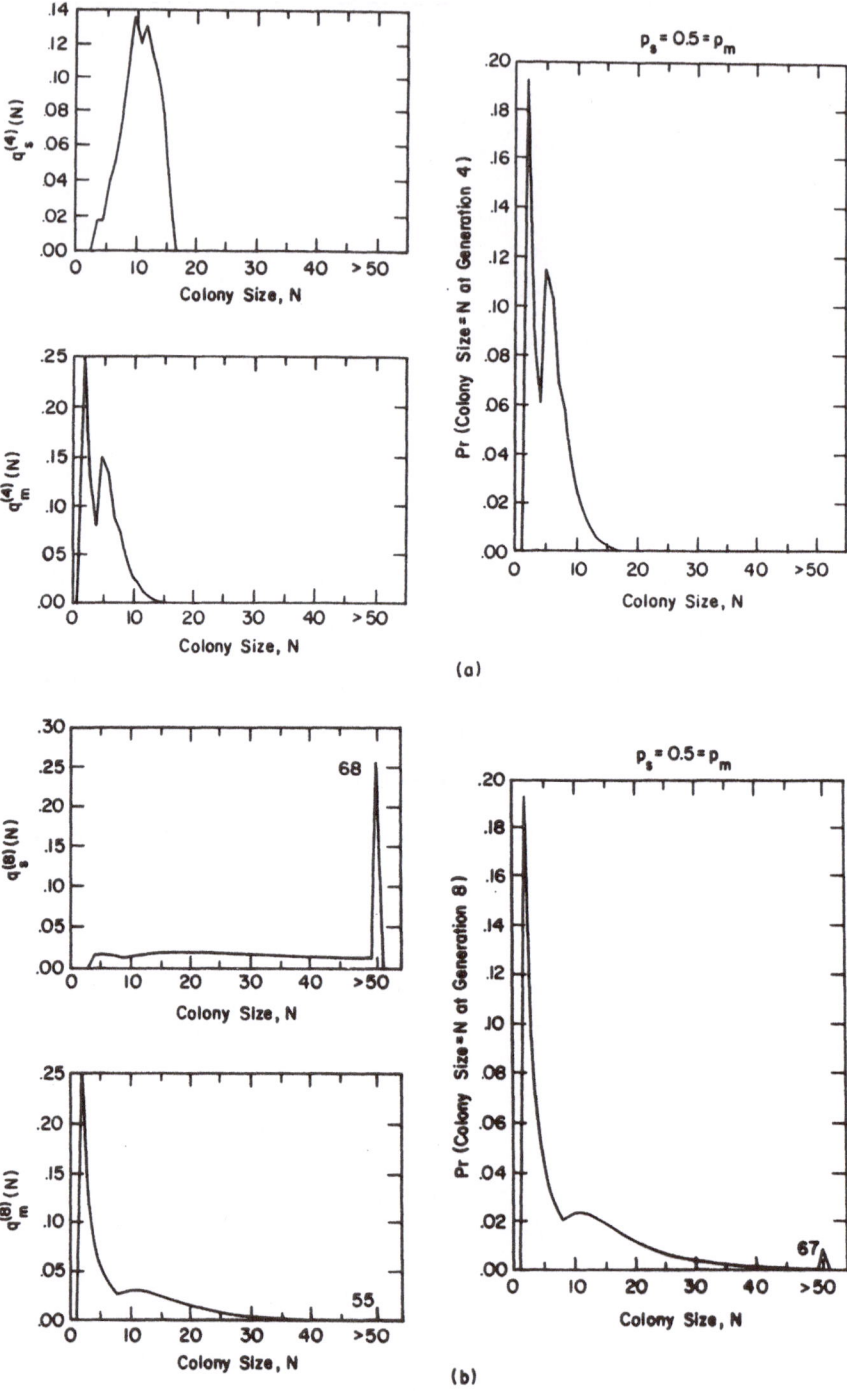

(a)

(b)

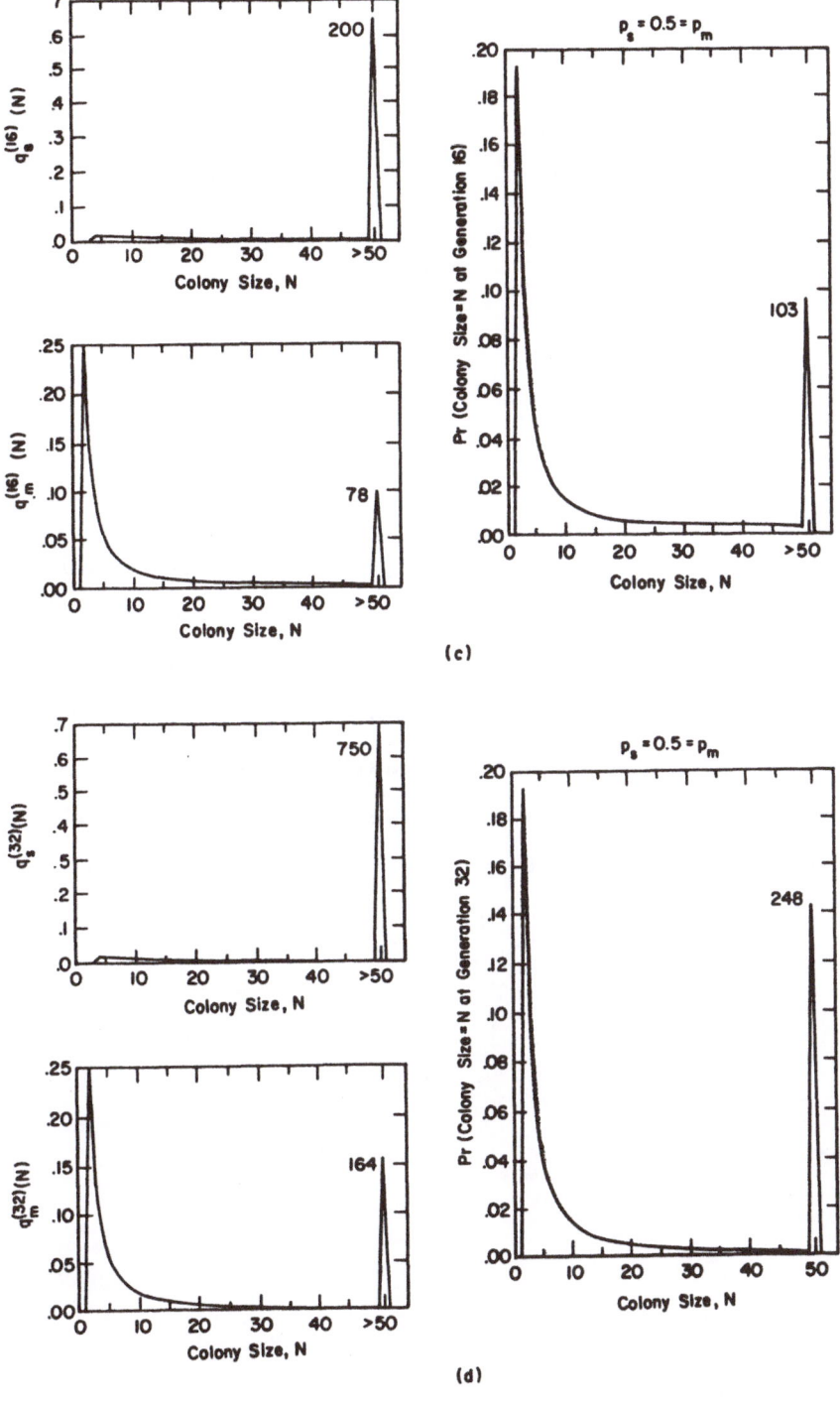

(c)

(d)

Figure 7.2 Probability that colony size $= N$, at generation n for colonies with S-cell, M-cell, or randomly chosen parent; (a) n $=$ 4, (b) n $=$ 8, (c) n $=$ 16, (d) n $=$ 32. Illustrated for $p_s = 0.6, p_m = 0.45$. Probability distribution for colonies with a random parent is a weighted average of the distributions for the known parents, where weighting is according to the fraction of cells of each type in the plated cell suspension. The upper tail of the distribution $(N > 50)$ is combined into a single category. The mean size of colonies in this category is written beside the probability peak.

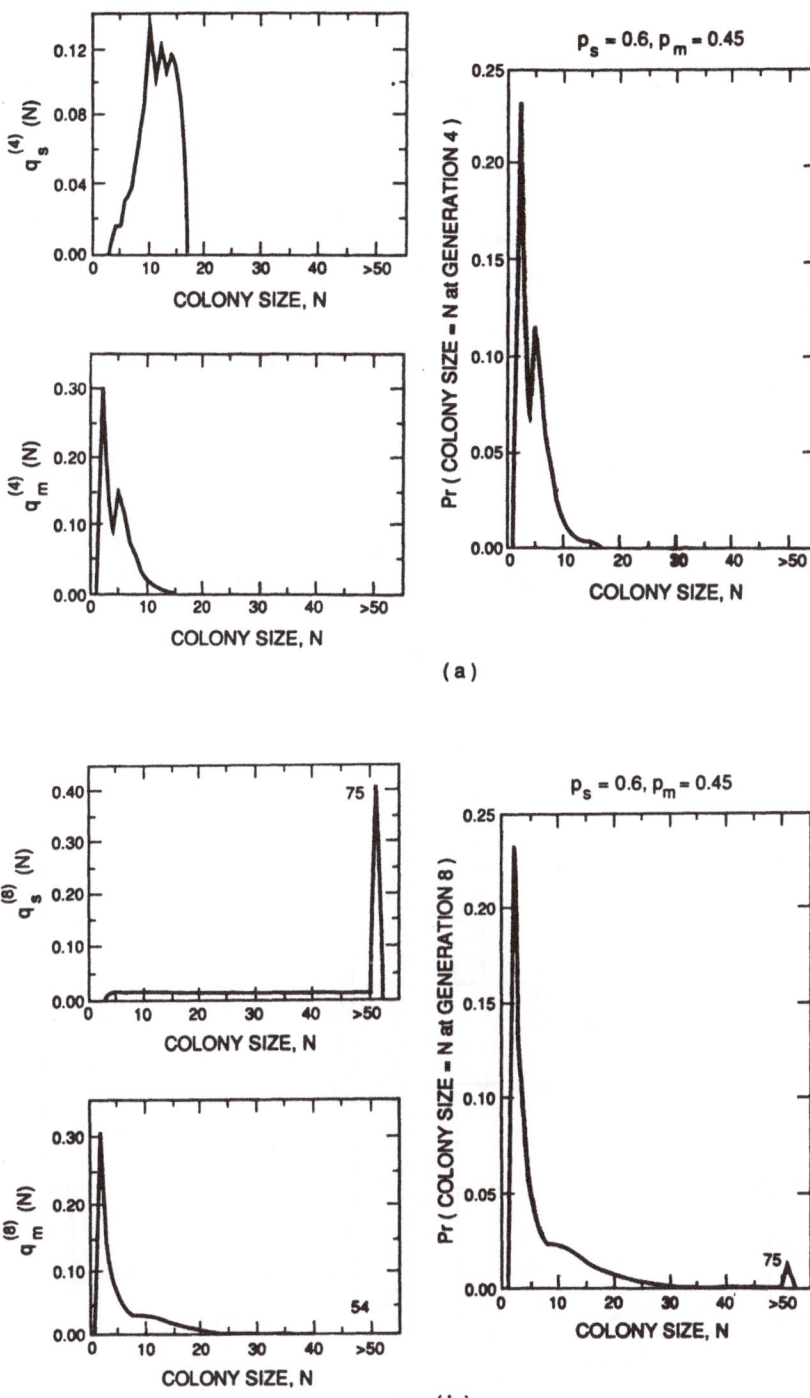

(a)

(b)

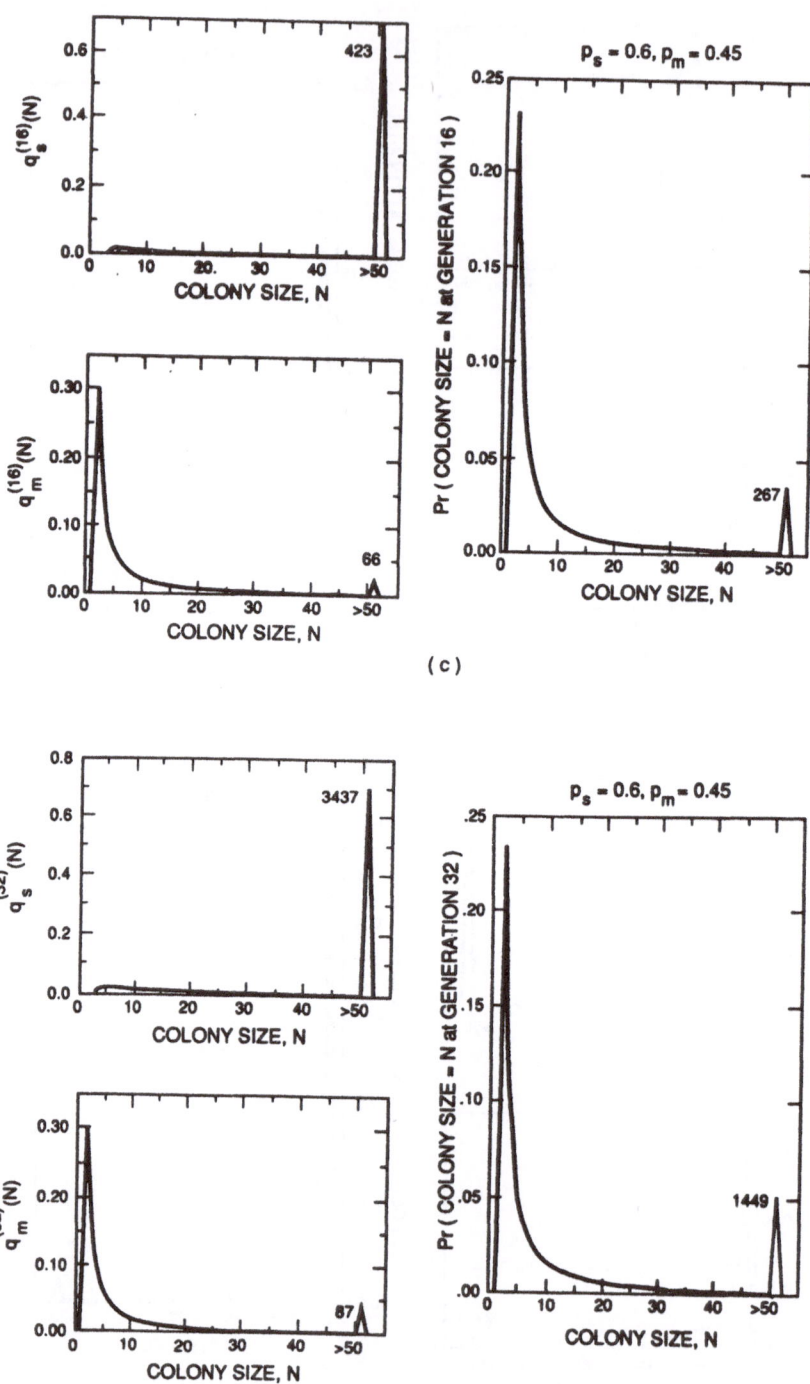

(c)

(d)

probabilities. However, the quantitative behavior of the graphs differed substantially between the two sets of parameter values, particularly at later times. Two important characteristics of colony growth, which we used to compare the sets of parameter values, were the probability that a colony would be "large" and the mean "large" colony size. In Table I, we present numerical values of these characteristics for each of the two cases. Clearly, colonies with a stem cell parent are much more likely to become "large" (i.e. > 50 cells) than colonies with a macrophage progenitor parent. In fact, after 32 generations, the most frequently observed colonies from stem cell parents are large, while the most frequently observed colonies from macrophage progenitor parents are small.

7.2.1 *Probability of a "large" colony*

In analyzing experiments, we are concerned with colony growth from a plate of mixed cell types. Thus in Figs. 7.1 and 7.2 we also plot the distribution of colony sizes that would arise from a mixture of parents of each type in the proportion η_s, η_m, η_e. By referring to Eq. (2.18), we can obtain

$$Pr(\text{colony size} = N \text{ by generation } n)$$

$$= \sum_{\alpha=S,M,E} \eta_\alpha Pr(\text{colony size} = N \text{ by generation } n | \text{parent of type } \alpha) \quad .$$

The probability that a colony is "large" (i.e., contains in excess of 50 cells) is obtained by accumulating $Pr(\text{colony size } = N \text{ by generation n})$ for $N > 50$.

7.2.2 *Mean size of a "large" colony*

To obtain the expected size of a large colony, i.e., $E(\text{colony size by generation } n | N > 50)$, we calculate a weighted average of the mean colony sizes for each of the types of parents, where the weighting is according to the relative likelihood of a large colony arising from a parent of each type. Hence, we calculate

TABLE I

Probability that a colony is large (> 50 cells) by generation n

(a) $p_s = 0.5 = p_m$

Generation, n	Parent		
	$\alpha = S$	$\alpha = M$	Mixture*
4	0(0)†	0(0)	0(0)
8	.26(68)	.001(55)	.01(67)
16	.65(200)	.10(78)	.10(103)
32	.70(750)	.16(164)	.14(248)

(b) $p_s = 0.6$, $p_m = 0.45$

Generation, n	Parent		
	$\alpha = S$	$\alpha = M$	Mixture*
4	0(0)	0(0)	0(0)
8	.41(75)	.00(54)	.01(75)
16	.70(423)	.02(66)	.04(267)
32	.71(3437)	.04(87)	.05(1449)

* Mixture indicates colonies started with either a stem cell or macrophage parent. For such colonies the probability that a colony is large is the weighted average of the probabilities for $\alpha = S$ and $\alpha = M$, with weights η_s and η_m, respectively, since end cells have probability zero of producing large colonies. The mean size of a colony is given by Eqs. (7.5) and (7.6) with $\eta_s = 0.03$ and $\eta_m = 0.77$.

† The mean size of a colony, given that it is large, is shown in parentheses.

$$E(\text{colony size by generation } n | N > 50)$$

$$= \sum_{\alpha=S,M} E(\text{colony size by generation } n | N > 50, \text{ parent of type } \alpha) \qquad (7.5)$$

$$\times \quad Pr(\text{colony has parent of type } \alpha | N > 50) \quad .$$

A colony with an end cell parent has size one and hence is excluded from the sum in Eq. (7.5).

7.2.3 Enrichment for stem cells

Because large colonies are more likely to arise from stem cell parents than from macrophage progenitor parents, we can expect that $Pr(\text{parent of type } \alpha | N > 50) \neq \eta_\alpha$, where η_α is the proportion of cells of type α in the original plate ($\alpha = S, M, E$). In fact, if we use Bayes Theorem (cf. Mood, Graybill, and Boes, 1974), we obtain

$$Pr(\text{colony has parent of type } S | N > 50) = \frac{\eta_s Pr(N > 50 | Z_0 = S)}{\sum_{\alpha=S,M} \eta_\alpha Pr(N > 50 | Z_0 = \alpha)} \qquad (7.6)$$

and

$$Pr(\text{colony has parent of type } M | N > 50) =$$

$$\qquad (7.7)$$

$$1 - Pr(\text{colony has parent of type } S | N > 50) \quad .$$

For example, when $p_s = 0.6$ and $p_m = 0.45$, then after 16 generations we find from Eq. (7.6) and Table I that $Pr(\text{colony has parent of type } S | N > 50) = 0.562$. The results of similar calculations for different parents and the two sets of values for the branching probabilities are summarized in Table II.

TABLE II

Enrichment for stem cell parent in large colonies

	Pr (parent of type α| colony size > 50)*			
	$p_s = 0.5 = p_m$		$p_s = 0.6,\ p_m = 0.45$	
Generation	$\alpha = S$	$\alpha = M$	$\alpha = S$	$\alpha = M$
8	0.93	0.07	0.997	0.003
16	0.20	0.80	0.56	0.44
32	0.15	0.85	0.41	0.59

* Entries in Table I are rounded. Table II was calculated using more significant digits than those quoted in Table I. These differences in significance lead to some entries in Table II differing from the values calculated using Table I directly.

From Table II, we see that the probability that a large colony arose from an S parent, and may therefore still contain S cells, decreases with time. Hence, a practical policy to implement if one were interested in enriching a culture for stem cells would be to continually replate the early-appearing large colonies. Further information on the prospects of enrichment for stem cells can be obtained by examining the finite-time probability distribution of the number of stem cells, calculated in Section 3.3. Caution must be exercised though to ensure that the size of the stem cell compartment is viewed in relation to the overall size of the colony. Even when the stem cell compartment is increasing in size, the macrophage and end cell compartments will be growing much faster and thus dominating the colony composition. (Refer also to Section 5.2.)

7.2.4 Bimodal colony size distribution

We predict from Table I that in plating out mixed cell types, we see an initial predominance of small colonies, but as time increases, some of the colonies (14% at generation 32 when $p_s = 0.5 = p_m$) continue to grow and become extremely large (corresponding mean = 248) while the remaining colonies (86% at generation 32 when $p_s = 0.5 = p_m$) are small, i.e., contain less than 50 cells. Thus, colonies tend to dichotomize into "large" or "small" categories. From Table I, we observe that the discrepancy between "large" and "small" colonies is vastly increased by generation 32 when we change to $p_s = 0.6$, $p_m = 0.45$. The average size (1449) of large colonies in this case is approximately five times greater than the corresponding average (248) when $p_s = 0.5 = p_m$.

7.2.5 Statistics of colony growth

The summary statistics of colony growth in Figs. 7.3 and 7.4 clearly demonstrate that one can expect enormous variability in the size of colonies for both types of proliferating parents. The standard deviation is much greater than the mean: in fact, from Eqs. (A.21) and (A.22), *standard deviation* $\cong n^{\frac{1}{2}} \times$ *mean*, where $n =$ generation, for the case $p_s = 0.5 = p_m$. Because the distribution of colony sizes has a very long tail, we cannot easily make a statement about the probability that a colony size will lie within one standard deviation of the mean for this distribution. However, the Winsor principle (cf. Tukey, 1960; p. 457) states that most distributions are normal in their middle portions. Therefore, by analogy with the normal distribution we may reasonably expect this probability to be around 60–70%.

7.2.6 Long-term behavior

Figure 7.4a shows that when colony growth is supercritical ($p_s = 0.6$, $p_m = 0.45$), the long-term limiting proportions in each of the cell types, predicted in Chapter 5, is attained quite quickly (after approximately 12 generations for colonies initiated by a stem cell or 8

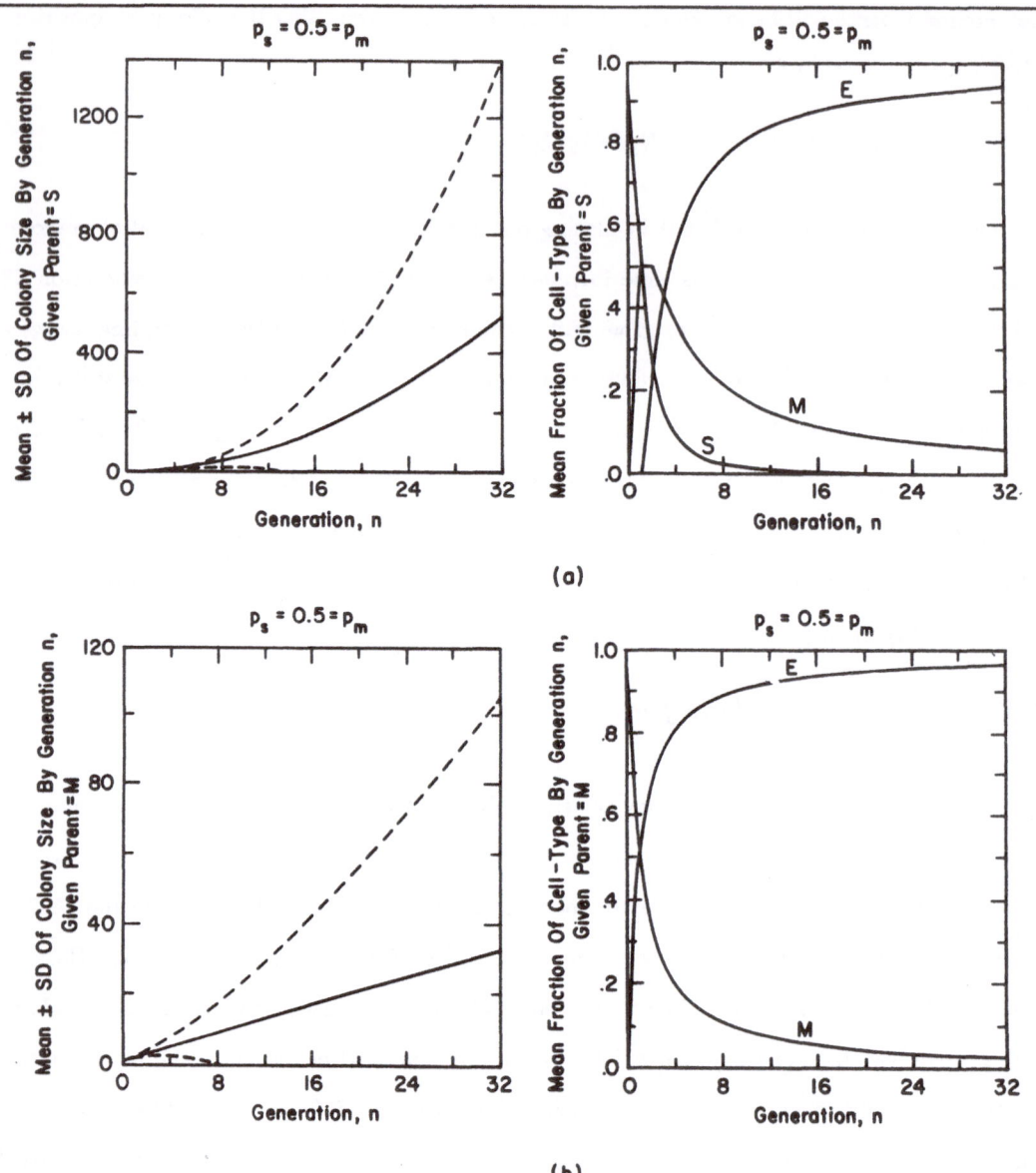

(a)

(b)

Figure 7.3 Statistics on colony growth as a function of time, for colonies with $p_s = 0.5 = p_m$. (a) Stem cell parent; (b) macrophage progenitor parent. The first frame of each pair gives *mean ± standard deviation* of the total colony size (where the lower bound is set to zero if it becomes negative). The second frame gives the expected number of each cell type as a fraction of the expected total colony size.

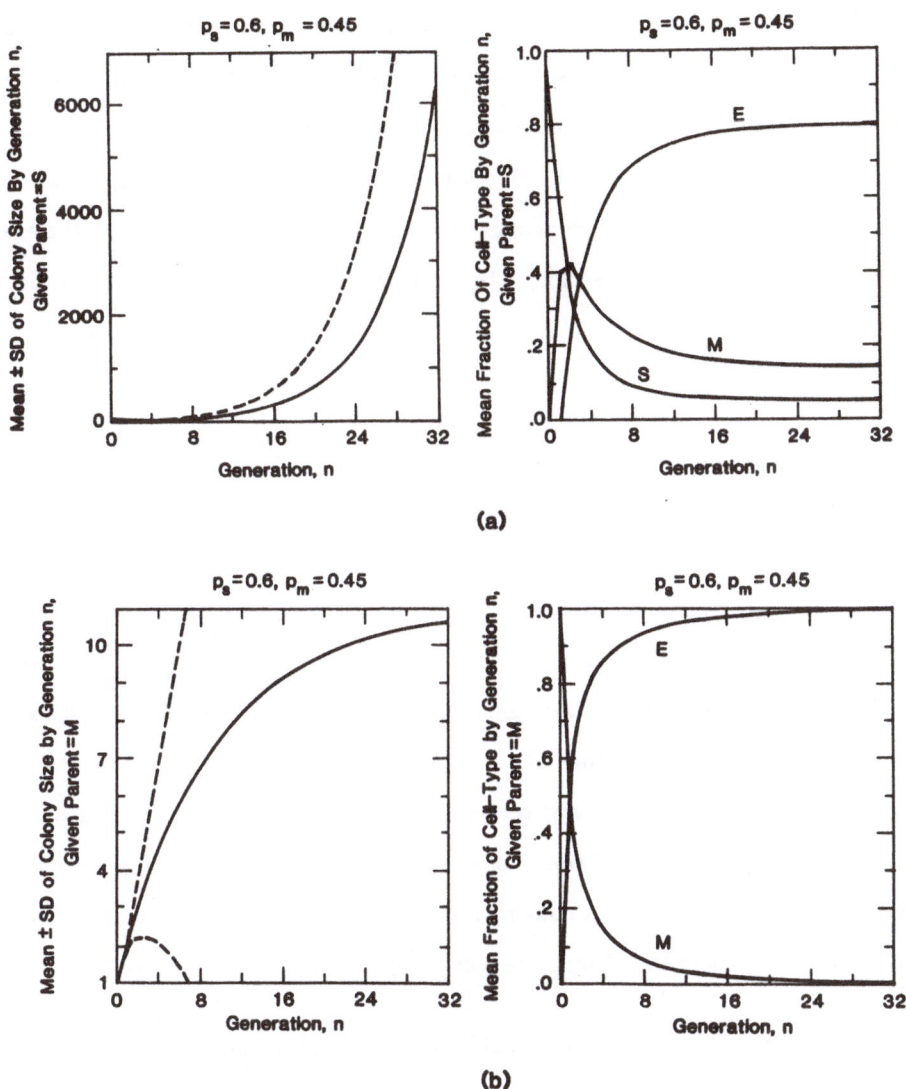

Figure 7.4 Statistics on colony growth as a function of time, for colonies with $p_s = 0.6, p_m = 0.45$. (a) Stem cell parent; (b) macrophage progenitor parent. The first frame of each pair gives *mean ± standard deviation* of the total colony size (where the lower bound is set to zero if it becomes negative). The second frame gives the expected number of each cell type as a fraction of the expected total colony size.

generations for colonies initiated by a macrophage). Further, in the critical and supercritical cases, as the number of generations increases, the colonies tend on average to be heavily dominated by end cells. This latter observation has implications for subculturing experiments: Even though a colony may be large and appear to be still growing, the expected proportion of stem cells is very small. Hence, on replating a large colony into several smaller subcolonies, there is a high probability that a subcolony contains no stem cells and will therefore be unsuccessful. Subcloning only a fraction of the cells in a colony will exacerbate the problem.

7.3 Completion of Growth

The rate at which colonies cease growing is observable experimentally in only rather broad terms since colony growth is rarely followed in detail over time. However, our model predicts that when a mixture of cells is plated out and allowed to grow, many, but not all, of the colonies will complete growth very quickly. When $p_m = 0.45$ (Fig. 7.5a), 50% of colonies initiated by a macrophage progenitor (the predominant initiating cell type) will have completed growth by about the second generation, leading to colonies of at most four cells. However some colonies keep growing. By 32 generations, close to 99% of colonies initiated by a macrophage progenitor will have ceased growing; thus 1% continue to grow. When $p_m = 0.5$ (Fig. 7.5b), about 10% of the colonies initiated by macrophage progenitors will still be growing after 32 generations. In contrast, for the case $p_s = 0.5 = p_m$, only slightly more than 50% of the colonies initiated by a stem cell will have ceased growing by that time.

If we allow stem cells to have a probability of self-replication closer to that which we believe is realistic ($p_s = 0.6$), we find that the rate of completion of growth of colonies initiated by a stem cell drops considerably. At 32 generations, when $p_s = 0.6$, $p_m = 0.45$, 41% of those colonies initiated by a stem cell will have ceased growing (Fig. 7.5a). Of course, stem cells constitute a very minor proportion of the mixture of cell types originally plated out. Hence, after 32 generations, a proliferating colony will be quite rare. In fact, for $\eta_s = 0.03$ and $p_s = 0.6$, $p_m - 0.45$, the probability that a colony is still proliferating after 32 generations is

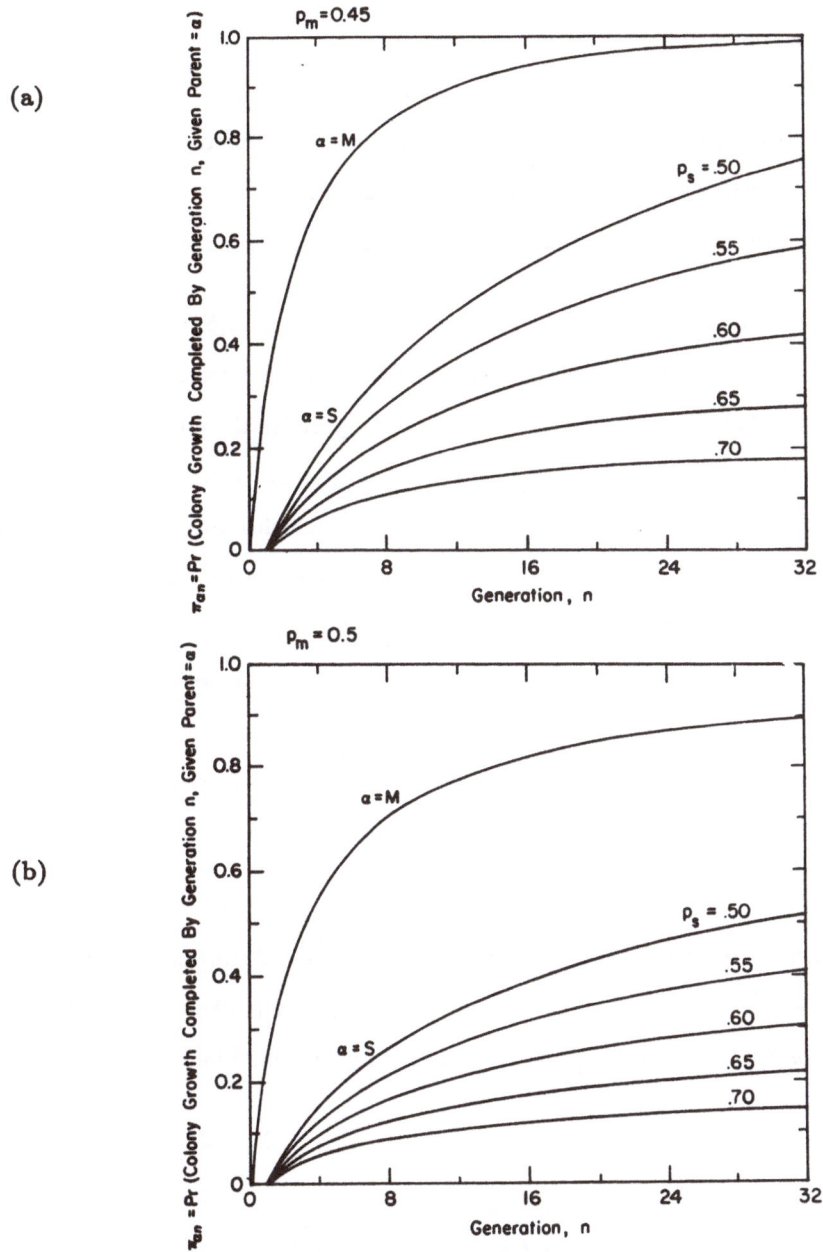

Figure 7.5 Cumulative probability distribution for number of generations to completion of colony growth when colonies have either a macrophage progenitor or stem cell parent. (a) $p_m = 0.45, 0.5 \leq p_s \leq 0.7$; (b) $p_m = 0.5, 0.5 \leq p_s \leq 0.7$.

approximately 0.025 (i.e., the fraction of stem cells plated, .03, times the probability that at 32 generations a colony initiated by a stem cell is still growing, .6, plus the fraction of macrophages plated, .77, times the probability that at 32 generations a colony initiated by a macrophage is still growing, .01). Among those colonies still proliferating, approximately 70% will have been initiated by a stem cell ($0.03 \times 0.59/0.025$). The results of Chapter 5 allow a prediction of the proportion of stem cells in these growing colonies.

It is relevant to ask about the asymptotes to the lines of Fig. 7.5; i.e., the probability that a colony will eventually cease growing. Our model predicts that certain ranges of values for the branching probabilities will lead to so-called "supercritical" growth in which the probability is less than one that the colony will eventually cease growing (Fig. 7.6). For $p_s = 0.6$ and $p_m = 0.45$, there is about a 55% chance that colonies initiated by a stem cell will continue to grow indefinitely, while colonies initiated by a macrophage progenitor will all eventually cease growing. (In fact, from Fig. 7.5a, about 99% of them will have ceased growing by generation 32.) Figure 7.6 shows that the probability of eventual completion of growth is rather sensitive to the value of the branching probabilities above the critical value of 0.5. Thus to make realistic predictions, precise estimates of p_s are needed.

7.4 Size at Completion of Growth

The time taken to reach completion of growth (Section 7.3 above) and the final size of the colony are related—the longer a colony grows, the larger it is likely to be. The exact correlation between final size and the time to completion is reflected in

$$Pr(Z_{sn} = 0, Z_{mn} = 0, Z_{en} = k | Z_0 = \alpha) \ , \quad \text{for } k = 1, 2, \ldots, 2^n \text{ and } n = 1, 2, \ldots \quad .$$

We calculate these quantities for $n = 2, 3, 4$ and $\alpha = S, M$. (Calculations of the probability distribution for larger finite n were impractical.) To reflect experimental observations, we take a weighted average over the probabilities, conditioned on the type of parent, to obtain

$$Pr(\mathbf{Z}_n = (0,0,k)) = \frac{\eta_s}{\eta_s + \eta_m} Pr(\mathbf{Z}_n = (0,0,k)|\alpha = S)$$

$$+ \frac{\eta_m}{\eta_s + \eta_m} Pr(\mathbf{Z}_n = (0,0,k)|\alpha = M) . \qquad (7.8)$$

Note that we have excluded from our calculations colonies of size one, for which $\alpha = E$. We exclude these colonies so that our calculations focus on the non-trivial growing portion of the clonal growth experiment. Finally, if one wishes to focus on only those colonies whose growth has ceased, one can normalize by dividing $Pr(\mathbf{Z}_n = (0,0,k))$ by $\sum_{k=0}^{\infty} Pr(\mathbf{Z}_n = (0,0,k))$. For example, of those colonies that have ceased growing by generation 4, the probability of containing at most two cells is $0.29/0.64 = 0.45$ when $p_s = 0.6$ and $p_m = 0.45$. Although we

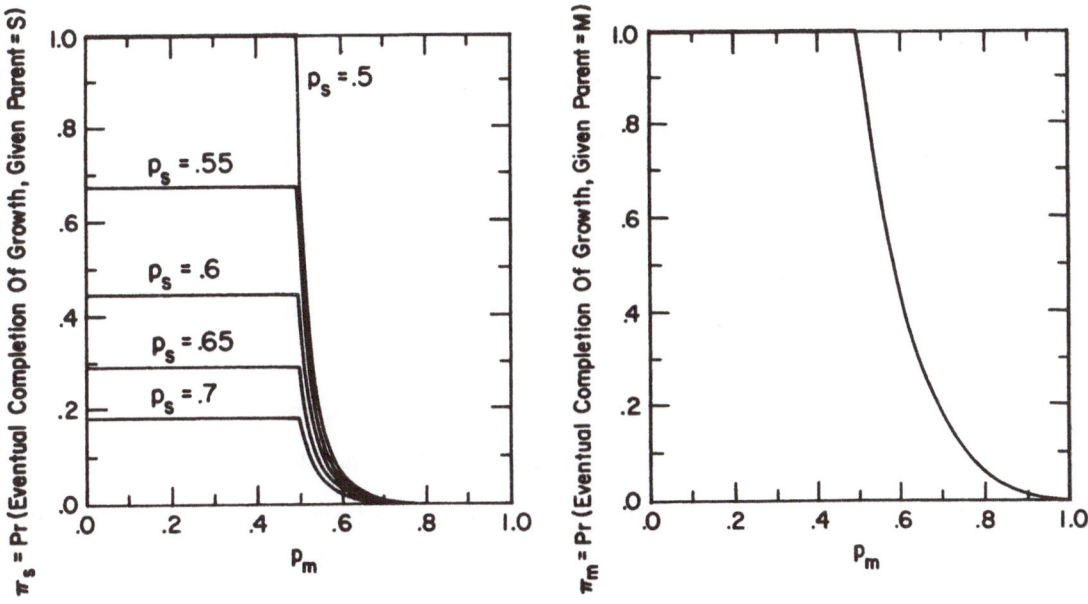

Figure 7.6 Limiting probability of completion of colony growth for colonies with either a stem cell parent or a macrophage progenitor parent. This figure provides the asymptotes to the families of curves in Fig. 7.5.

have not explicitly included this normalization, the results of Table III reinforce the notion that many of the colonies stop growing very quickly and are rather small. Graphs of the probability distribution of the eventual (i.e., infinite time) final colony size are rather easier to obtain, via the simple recursions of Eqs. (4.11) and (4.13) (see Fig. 7.7). These graphs are potentially useful for comparing with data. As seen above, many of the colonies are predicted to cease growing rather quickly, and hence, after a period of time of, say, 32 (or even 16) generations, one expects the majority of the colonies to have reached their final size. Our model predicts that colonies produced by macrophage progenitor parents, even when growth is critical (i.e., even when $p_m = 0.5$), are likely to be small and to fall into the "cluster" category. Reinforcing results that we have seen in the earlier graphs, we see that colonies produced by stem cell parents are likely to be much larger than those of macrophage progenitor parents. The final size distribution of colonies with stem cell parents is particularly sensitive to changes in the value of p_m. This is because the proliferation in the stem cell compartment magnifies the effects of change in the subsequent compartments in the model.

7.5 Asymptotic Proportion of Each Cell Type

Under conditions that give a positive probability of growing without bound, we examine the limiting proportion in each of the three compartments of the model [see Sections 5.1 and 5.2]. Table IV gives the limiting proportions for values of p_s and p_m near their critical values, and for p_s or p_m near one. (As a reminder, in our model, colonies initiated by a macrophage can never contain stem cells.) For the particular choice of $p_s = 0.6$, $p_m = 0.45$, Figure 7.4a shows that on average the limiting proportion is closely approximated after only 12 generations, thus giving some indication of the magnitude of "long term" in the context of clonal growth. Notice that in realistic cases, end cells predominate in the colony. Since *in vivo* these functionally mature cells would have left the bone marrow, we do not observe this high a proportion of end cells in plating out samples of bone marrow. However, the proportion of stem cells in long-term colonies is approximately the same as that in bone marrow. In fact, $p_s = 0.57$ and $p_m = 0.45$ gives the proportion of stem cells equal to 0.03.

TABLE III

Probability that a colony completes growth by generation n
and contains at most k cells [see Eq. (7.8)].

(a) $p_s = 0.5 = p_m$

$n = 2$		$n = 3$		$n = 4$	
k	Probability	k	Probability	k	Probability
2	.24	2	.24	2	.24
3	.36	4	.44	4	.44
4	.38	6	.47	6	.52
		8	.47	8	.53
				16	.53

(b) $p_s = 0.6$, $p_m = 0.45$

$n = 2$		$n = 3$		$n = 4$	
k	Probability	k	Probability	k	Probability
2	.29	2	.29	2	.29
3	.44	4	.52	4	.52
4	.45	6	.56	6	.62
		8	.56	8	.63
				16	.64

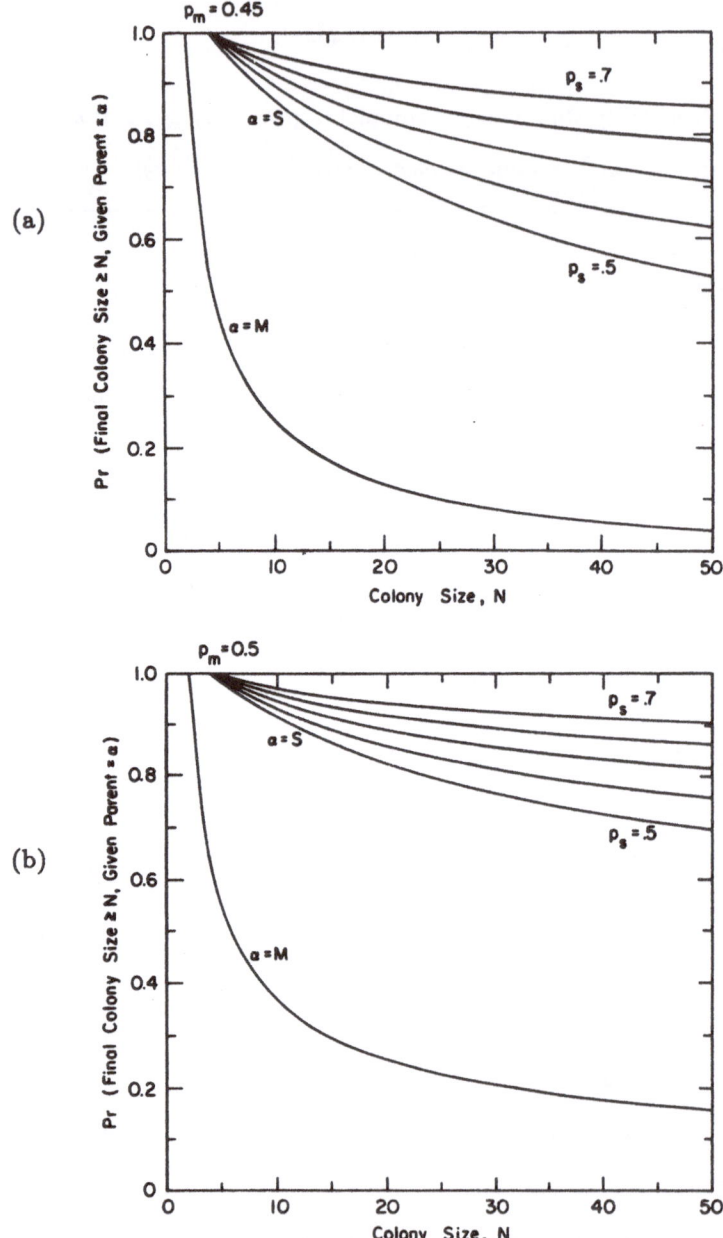

Figure 7.7 Probability distribution of colony size at completion for colonies with a stem cell or macrophage progenitor parent. (a) $p_m = 0.45, 0.5 \leq p_s \leq 0.7$; (b) $p_m = 0.5, 0.5 \leq p_s \leq 0.7$.

TABLE IV

Limiting proportion of cells of each type (S, M, E) as a function
of branching probabilities, p_s, p_m, for colonies with:

(a) S-cell parent

p_s

p_m	0.55	0.60	0.90
0.45	(.02, .08, .90)	(.05, .15, .80)	(.65, .15, .20)
0.50	(.01, .09, .90)	(.04, .16, .80)	(.64, .16, .20)
0.55	*	(.02, .18, .80)	(.62, .18, .20)
0.90	(0, .90, .10)	(0, .90, .10)	*

*   When $p_s = p_m$, a separate complex derivation of the limiting
proportions is required. Since this case is rare in practice, calculations
were omitted.

(b) M-cell parent

p_m	0.55	0.60	0.90
	(0, .10, .90)	(0, .20, .80)	(0, .80, .20)

Chapter 8

Conclusions and Extensions

Since the demonstration of the existence of a pluripotent hemopoietic stem cell (cf. Till et al., 1964), there has been discussion about the nature of the commitment and differentiation processes that cause the undifferentiated stem cell to develop into a functionally mature hemopoietic cell. Two opposing views of the process have developed: one view proposes that the decision of when, and along what pathway, a stem cell differentiates is preprogrammed in the cell, i.e., that decisions are under strict genetic control. Hence, offspring of a stem cell will pass through a regulated sequence of transformations until the (predetermined) end point is reached. This view we call *deterministic*. The other view, which we call *stochastic*, allows for the commitment process to be random. In the stochastic view a decision to commit may be made in any cell cycle. However, not all pathways may be available to a progenitor at a given degree of maturity. Thus the cell is only able to commit, with certain (unknown) probabilities, to any one of the pathways available. Furthermore, the probability of commitment may not be fixed but rather may be a function of the cell's environment and the cell's biochemical state. In particular, one may envision the situation in which the local concentration of hemopoietic growth and differentiation factors (e.g., CSF-1, IL-3, IL-1) and the number of cell surface receptors for these growth factors affect the probability of commitment.

One may expect that the level of control implied by the deterministic model of commitment would result in a large degree of synchrony in the outcome of clonal growth experiments. Such synchrony is not observed. Quite the contrary, clonal growth patterns are often characterized by their extreme variability, both in the size and in the composition of the clones (cf. Pharr et al., 1984; Nakahata and Ogawa, 1982; Suda et al., 1983a). Such high variability has been used as support for the appropriateness of a stochastic model; we ourselves appeal to this observed variability in our selection of a stochastic model.

The model that we developed in this text was couched in terms of macrophage differentiation. We attempted to construct the simplest stochastic model that retained what we considered to be the key cell populations in any differentiation pathway: stem cells (S), proliferating (macrophage) progenitor cells (M), and fully differentiated end cells (E). The approach that we took to describe self-renewal is also very simple – we assumed that the offspring of a proliferating cell (S or M) have stochastically independent fates. That is, we assumed that with probability p_s an offspring of a stem cell remains a stem cell, and with probability p_m an offspring of a macrophage progenitor remains a macrophage progenitor. We chose to ignore the possibility of cell death. Thus, in order for stem cells to self-renew, we chose $p_s \geq 0.5$. Because progenitors do not self-renew, we chose $p_m \leq 0.5$. The case of $p_s = p_m = 0.5$, although perhaps not realistic, is a "critical case" that we studied to gain mathematical insight into our model.

Our model is a discrete time model that follows cells from one cell division to the next. In our model, commitment is made some time between cell divisions resulting in the offspring of cells of one type becoming cells of another type. There is evidence that the decision to commit may be made at the level of the cell's genome before there is any evidence of commitment at the phenotypic level. The work of von Melchner and Höffken (1985), Steinberg and Brownstein (1982), and Gusella et al. (1976) indicates that at a random time during the cell cycle a biochemical event may occur, possibly involving reorganization of the genome or expression of certain genes, which sets in motion a series of further events, in some cases occurring over several cycles, that leads to the development of a differentiated phenotype. Our model does not attempt to go into mechanistic detail; rather, we simply assume that a stem cell becomes a progenitor cell when its progeny have a decreased probability of renewal.

The fundamental parameters of our model are p_s and p_m. At a phenomenological level we might express the probabilities p_s and p_m as functions of environmental parameters, thus allowing, for example, the probability to commit at a particular time to be correlated with the concentration of growth factors present. If formal procedures were used for estimation of these probabilities, then it might be possible to test for the role of the environmental factors on the probability to commit.

For the sake of simplicity, we assume that the fate of a particular cell is independent of the fates of the other cells in a colony. Thus, for example, our model can not allow intercellular communication or density-dependent growth. [Density dependent branching processes have been studied by Fujimagari (1976) and Klebaner (1983, 1984) among others.] We also cannot include "aging" within a cell line, where age is the number of divisions since the initial parent of the cell type. Density dependence can be controlled in experiments by subculturing a colony when it gets too large. Intercellular communication undoubtedly exists, but its effect in the current context have not been measured. The study of cell aging or cell senescence is a field by itself (cf. Hayflick and Moorhead, 1961; Hayflick, 1965, 1979; Kirkwood and Holliday, 1975; Holliday et al., 1981) and is outside the scope of this work.

The assumption of independence of fates of daughter cells is quite strong and needs to be tested. There does, however, appear to be clear evidence that cell division is asymmetric. The model that we developed here can be easily adjusted to account for correlation between the fates of daughter cells. Of course, numerical predictions would then change.

Although we have assumed that the S and M compartments are homogeneous, we realize that this is only a rough approximation to the truth: the macrophage progenitor compartment in fact comprises several identifiable compartments in series, each with its own characteristic branching probability; evidence for heterogeneity in the stem cell compartment also exists (cf. Hodgson and Bradley, 1979; Burgess and Nicola, 1983) although the causes for this heterogeneity are not known. There is insufficient experimental data available, however, to warrant modeling heterogeneity within the compartments, and hence we opted for simplicity and applied the same branching probabilities to all cells within a given compartment.

The time unit that we have adopted in our discrete time model is a cell generation time. In order for our results to be interpretable in "real" time, a generation should correspond to a fixed, constant cell cycle time for all cells and all cell types in the population. We have assumed that this correspondence exists. We have also implicitly assumed that the growth fraction is one and hence that all cells are in cycle. Further, we have assumed that when a population of cells is placed on a plate, all of the cells divide in synchrony. These assumptions are idealizations. Synchrony can be obtained in experiments but generally it is not. All cells are not in cycle. In fact, for the stem cell compartment, the ability of cells to cycle very slowly, or even remain out of cycle, is likely to be an important factor in the regulation of hemopoiesis *in vivo*. Hemopoietic cells *in vivo* must be replenished constantly throughout life, thus making continual demands on the proliferative capacity of the stem cell pool. A vital question is: how is the population of stem cells protected from depletion due to, say, severe stress? It appears that only about 10% of stem cells are actively proliferating at any one time (Burgess and Nicola, 1983). Under demand for hemopoietic replenishment, up to 50% of stem cells can be in active cycle. However, the differentiation rate of actively proliferating stem cells may be constant and independent of the rate of proliferation. In the future it may be desirable to introduce into our model cell cycle times which differ among cell types, or even vary continuously about a mean value that is characteristic of a cell type.

To obtain insight into our model, we selected a set of branching probabilities ($p_s = 0.6$, $p_m = 0.45$) that we felt were realistic. With these values we made predictions on characteristics of colony growth such as changes in the size of colonies with time, the time until colonies completed growth, and the composition of the rare, very large colonies. When appropriate, we compared predictions using our "experimental" values of the branching probabilities with predictions using so-called "critical" values, namely $p_s = 0.5 = p_m$. The critical values mark the threshold between two modes of growth, subcritical (in which all colonies must eventually cease growing) and supercritical (in which colonies have a non-zero probability of growing without bound). Many of the predictions about colony growth show that the multiplicative nature of the branching process induces great sensitivity to small changes in p_s.

The most obvious characteristic of our predicted colony growth was the extreme variability under both sets of parameter values in the total colony size with time. In fact, the standard deviation of colony size was on the order of $\sqrt{n}$ times the mean colony size in generation n. Thus, individual colony size alone is a poor measure for comparison of self-renewal capacity of progenitors. Full cognizance must be taken of the large variability expected among colonies produced by progenitors with identical self-renewal capacity.

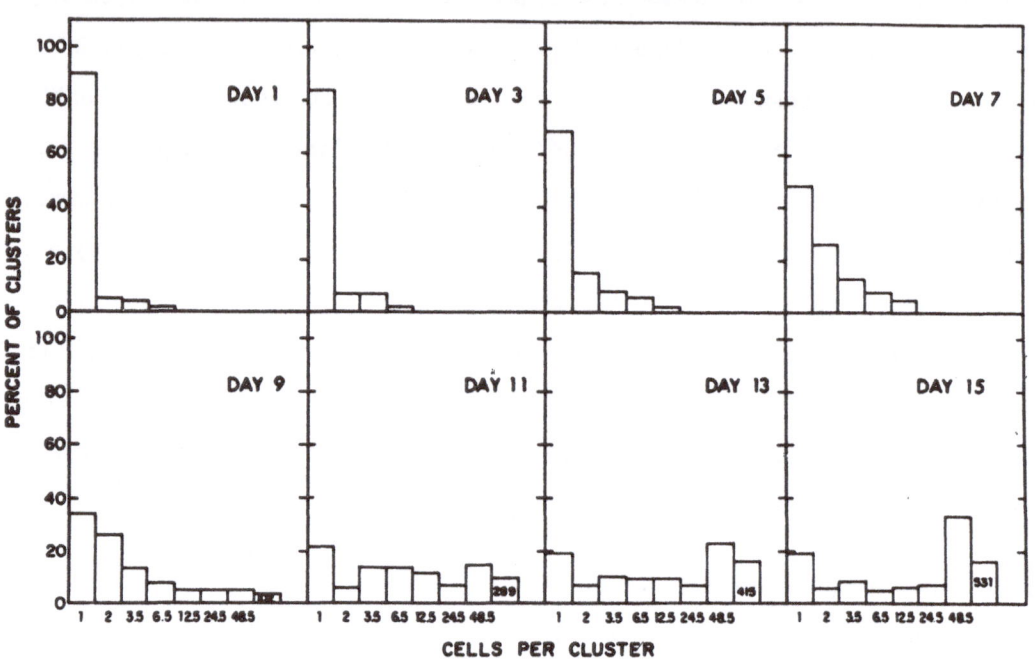

Figure 8.1 Histogram of the relative frequency of clusters of different sizes formed by mature mononuclear phagocytes. Taken from Stewart (1980). Size categories are: 0-1, 2, 3-4, 5-8, 9-16, 17-32, 33-64, > 64; categories are labeled by their midpoint, where appropriate. The number in the > 64 bar gives the mean size of colonies in this category.

One aspect of the variability in colony size with time, bimodality, was not apparent until a number of cell cycles had occurred. Under the conditions given in Fig. 7.1 at least eight cell cycles were needed. This was the earliest time for which colonies appeared to segregate into two classes – large (≫ 50 cells/colony) or small (≤ 50 cells/colony). At later times, for example, by generation 16, the segregation was more apparent, with 10% of colonies in the "large" category when $p_s = 0.6$, $p_m = 0.45$. Bimodality is not an artifact of the method of data presentation because $p_m < 0.5$ implies that eventually (in fact, quickly) all colonies with macrophage parents, even the so-called "large" colonies, will cease growing. In order that the "large" category contains colonies that continue to increase in size with time, stem cells having $p_s > 0.5$ must be present. Predictions of the time until a colony ceased growing shed light on this phenomenon. Macrophage progenitors, which were the predominant cell type in the population of cultured cells, formed colonies that ceased growing very quickly (for example, 50% had completed growth by the fourth generation even when $p_m = 0.5$) and were, not surprisingly, small. On the other hand, some of the rare stem cells continued to divide, producing very large colonies indeed. (After 32 generations the mean colony size from stem cell parents with $p_s = 0.6$, $p_m = 0.45$ was 3437.) Our prediction of bimodality in the colony size distribution loosely mimics the bimodality observed by Stewart (1984) in his experiments on the growth of murine myelomonocytic cells (Fig. 8.1). One apparent discrepancy between predictions and experimentation is that the experimental observations lag in time. The histograms depicting cell growth are essentially the same for days 1, 2, and 5, suggesting that few cells proliferate in the early stages of the experiment. Eventually, experimental bimodality appears after nine days, which, with a cell cycle time of approximately 16 hours, translates to 13 generations. The discrepancy between experimentation and theory may be due to the fact that stem cells in the culture are quiescent, or cycling very slowly, and require a few days in culture before the majority of them enter active cell cycle. It would not be difficult to generalize our model to allow for this phenomenon. For example, we could precede our stem cell compartment with a compartment of quiescent (G_0) cells having a random exit time after addition of stimulating growth factor (see also Suda et al., 1983b). C. Stewart (personal communication) suggested as another explanation that end cells may produce a factor that inhibits proliferation of stem cells, which is ultimately overcome by the presence of added growth factors.

In light of the preceding discussion, we have confidence in our model since important characteristics of observed colony growth, such as extreme variability in the size of colonies and bimodality in the colony size distribution, are qualitatively matched by our model. In particular, our model allows us to study the stem cell compartment, which is difficult to examine experimentally.

A controversial aspect of the stem cell is its probability of self-renewal and, indeed, the meaning of self-renewal. Frequently, self-renewal is held to mean the long-term maintenance of a steady size of the stem cell population. The stem cell compartment is increased by self-replication and decreased by differentiation, the size of the compartment resulting from the balance between these two processes. In the long term, in order that the size remain constant, each stem cell must, on average, give rise to one stem cell, all other offspring being differentiated cells. Our model as it applies to *in vitro* experimentation assumes that, with probability p_s, an offspring of a stem cell remains a stem cell. Thus with probabilities $(1-p_s)^2$, $2p_s(1-p_s)$, and p_s^2, a stem cell, when it divides, produces zero, one, or two daughter cells that are exact replicas of itself, and the mean number of stem cells produced is $2p_s$. By examining the probability that a colony will complete growing, we determine that $p_s > 0.5$ ensures a positive probability of a stem cell line always containing at least one stem cell. In contrast, $p_s < 0.5$ ensures that all daughter cells of a stem cell line will eventually differentiate to become macrophage progenitors. Since it seems clear that macrophage progenitors have a limited self-renewal capacity (i.e., $p_m < 0.5$), continual replenishment of the stem cell compartment *in vitro* therefore depends on $p_s > 0.5$.

Our model leads to two important implications for *in vitro* experimentation. First, subculturing may lead to false conclusions. According to our model, $p_s > 0.5$ implies there is a positive probability of a stem cell colony growing without bound. A major limitation to observing extremely large colonies experimentally is the need to subculture when cells become crowded on a plate, in order that constant culture conditions can be maintained. Our model predicts that a colony that has grown for many generations (and is therefore likely to be very large) will contain 5% stem cells when $p_s = 0.6$ and $p_m = 0.45$. Thus, if only a fraction of a colony is subcultured, the likelihood of including a stem cell in the subculture may be very low (depending, of course, on the dilution factor). Without a stem cell, the culture will eventually fail to grow, thus creating the impression of a mortal cell line.

The second implication involves enrichment for stem cells. A major problem in experimentation on stem cells is their low rate of occurrence in bone marrow. Our model suggests a method of enriching for stem cells. Since we predict that stem cells are more likely to produce large colonies than are macrophage progenitors, selecting a large colony for replating should give a greater chance of obtaining a stem cell than would selecting a small colony. In Table II, we calculate the probability that a colony with $N > 50$ cells was initiated by a stem cell or a macrophage progenitor as a function of the number of generations of cell growth. Only those colonies initiated by a stem cell have a positive probability of still containing stem cells. Since $p_s > 0.5$, the stem cell line is expected to grow and has a low probability of becoming extinct in just a few generations. As Table II shows, the most effective enrichment strategy is to select large colonies after only 8 generations.

Branching processes provide a general framework for the quantitative study of a broad range of scenarios involving cell differentiation and proliferation. For example, a simple rewording of the theory of supercritical processes (as presented in Chapter 5) has implications for experimentation on the action of reported differentiation-inducing factors on leukemic cells. In many ways (e.g., morphologically and on the basis of receptor expression) leukemic cells appear to be analogous to normal progenitor cells but are cells that are arrested at an immature stage of development since they have the capacity for infinite self-renewal when incubated in medium alone, an environment in which normal cells will not proliferate. Recently, Souza et al. (1986) have shown that the human growth factor, granulocyte colony stimulating factor (G-CSF), which supports clonal expansion of granulocyte progenitors *in vitro*, can induce *in vitro* differentiation of certain classes of human acute nonlymphocytic leukemic cells into granulocytes and macrophages. However, experimentalists do not always observe that G-CSF is an effective differentiation-inducing agent. For example, J. D. Watson and associates (personal communication) conducted a series of experiments in which samples of bone marrow aspirated from patients with various classes of acute myeloid leukemia were incubated *in vitro* with medium containing G-CSF. In the majority of samples, and in particular in samples from patients with leukemia of the classes for which Souza et al. (1986) reported their successful experiments, no maturation of leukemic cells was observed. Three conflicting interpretations are possible. The first infers that no differentiation took place, i.e., that G-CSF was ineffective as a differentiation-inducing agent. The second infers that differentiation took place in a small proportion of the leukemic cells, but the mature offspring were undetectable because of the vast preponderance of other cells. The second inference suggests a sampling problem upon which our model can shed some light. The third infers the leukemic cells are in a lineage that is not sensitive to the differentiation-inducing effects of G-CSF, e.g., are committed to the macrophage lineage.

We relabel our three compartments L (leukemic), D (differentiated), and E (end) cells, where $L \rightarrow D \rightarrow E$. Similarly, we relabel the corresponding branching probabilities p_l, p_d, and p_e. If we suppose that a leukemic cell will, with probability one, replicate itself at each cell division in the absence of differentiation–inducing factor, then the model predicts exponential growth of the L compartment. Suppose a differentiation–inducing factor, such as G-CSF, causes a small proportion, δ, of the L compartment to differentiate. On a population level, this is equivalent to supposing that, in the presence of G-CSF, $p_l = 1 - \delta$, $(\delta > 0)$, and now we expect some contribution to the D and E compartments of the process. To fully specify our model, we must select values for p_d and p_e. We shall let $p_d = 0.45$ since the differentiated cells are expected to behave similarly to normal committed progenitor cells, and $p_e = 1$ as before. (This last assumption may be better if modified to allow for cell death in a time frame of 7-10 days, which is that of the experiments described.) For any value of $0 < \delta < 0.5$, we can use the results of Chapter 5 to predict the long-term proportion of cells of each type in the differentiation studies. If $\delta = .01$, the limiting proportion (L, D, E) is $(.96, .02, .02)$. Similarly, if $\delta = 0.05$ (0.10), the limiting proportion is $(.82, .08, .10)$ $[(.65, .15, .20)]$. Apparently, if a factor such as G-CSF does indeed induce differentiation, the effect may not be observable if it is small because leukemic cells will comprise the major proportion of the sample, a situation that will be aggravated if progenitors (D cells) are morphologically indistinguishable from leukemic cells and further aggravated if leukemic cells make up less than 100% of the initial sample of cells plated.

An important extension of our study that utilizes the branching process framework is the inclusion of progenitors intermediate between stem cells and macrophage progenitors. For example, the macrophage progenitor (M cell) is known to be a differentiated offspring of a bipotential granulocyte-macrophage progenitor (GM cell). Growth and differentiation of the GM cell is supported by granulocyte-macrophage stimulating factor (GM-CSF). A daughter cell of a GM progenitor may differentiate into either a macrophage progenitor (our M cell) or a granulocyte progenitor. Metcalf (1980) used clone-transfer experiments to demonstrate that offspring of a GM progenitor choose at random one of the two available single-lineage differentiation pathways. Further, the probability of choosing a particular pathway appeared to be influenced by the concentration of GM-CSF. The extension of our model to include bipotential progenitors heading two independent single-lineage pathways is obvious. Such a model could provide a powerful tool for examining the effects of purported differentiation-inducing agents.

We believe that our model, although admittedly simplistic, predicts the readily observable broad characteristics of colony growth – extreme variability in colony size and length of time before growth ceases – while adding valuable insights into the details of the vital stem cell compartment of the hemopoietic system. We have already mentioned possible extensions of the model to allow for heterogeneity in the probabilities of self-replication and to account for a possible latency period in the early culture times. As more is known about genes expressed during proliferation and differentiation, it will be possible to model the branching probabilities in detail, thus leading to a better understanding of the sources and degrees of control in the differentiation of the hemopoietic stem cell.

Appendix A

Moments of Z_n

Let $\mathbf{Z}_n = (Z_{sn}, Z_{mn}, Z_{en})$ be the population composition after n generations. In this appendix we shall compute the mean, $E[\mathbf{Z}_n|\mathbf{Z}_0]$, and variance, $var[\mathbf{Z}_n|\mathbf{Z}_0]$, of $\mathbf{Z}_n$ given an initial single cell as the parent of a colony. Our approach is to obtain a set of recursion relations involving the moments of interest, which we then use to develop a coupled system of equations involving the generating functions of the moments. This system of equations is solved for the generating functions, which then yield the desired moments as coefficients of polynomial expansions in a dummy variable.

A.1 Mean

We first consider the mean of the process. To obtain a set of recursive relationships involving $E[Z_{\beta n}|\mathbf{Z}_0]$, $\beta = S, M, E$, we take partial derivatives of Eq. (2.8) with respect to θ_β and evaluate at $\boldsymbol{\theta} = 1$. For example, to obtain $E[Z_{sn}|\mathbf{Z}_0]$, we calculate

$$\frac{\partial f_\alpha^{(n)}(\boldsymbol{\theta})}{\partial \theta_s}\bigg|_{\boldsymbol{\theta}=1} = \sum_{i\geq 0}\sum_{j\geq 0}\sum_{k\geq 0} i\, Pr(\mathbf{Z}_n = (i,j,k)|\mathbf{Z}_0 = \alpha)$$

$$= E\left[Z_{sn}|\mathbf{Z}_0 = \alpha\right] \quad, \quad (\alpha = S, M, E) \ . \tag{A.1}$$

The same process of partial differentiation and evaluation applied to the components of $\mathbf{f}$ in Eq. (2.9) gives, for example,

$$\frac{\partial f_\alpha^{(n)}(\boldsymbol{\theta})}{\partial \theta_s}\bigg|_{\boldsymbol{\theta}=1} = \frac{\partial f_\alpha^{(n-1)}(\boldsymbol{\theta})}{\partial \theta_s}\bigg|_{\boldsymbol{\theta}=1} \left[\frac{\partial f_s(\boldsymbol{\theta})}{\partial \theta_s} + \frac{\partial f_m(\boldsymbol{\theta})}{\partial \theta_s} + \frac{\partial f_e(\boldsymbol{\theta})}{\partial \theta_s}\right]\bigg|_{\boldsymbol{\theta}=1} \tag{A.2}$$

$$= 2p_s\, E\left[Z_{s,n-1}|\mathbf{Z}_0 = \alpha\right] \ ,$$

where we have used the fact that $\dfrac{\partial f_m(\boldsymbol{\theta})}{\partial \theta_s} = \dfrac{\partial f_e(\boldsymbol{\theta})}{\partial \theta_s} = 0$.

Combining Eqs. (A.1) and (A.2) we obtain the recursion relation,

$$E\left[Z_{sn}|Z_0 = \alpha\right] = 2p_s\, E\left[Z_{s,n-1}|Z_0 = \alpha\right] \ . \tag{A.3}$$

Similarly, expressions can be obtained for the remaining partial derivatives of Eqs. (2.8) and (2.9). If we use the matrix of means defined in Eq. (2.13), i.e.,

$$M = \begin{pmatrix} 2p_s & 2(1-p_s) & 0 \\ 0 & 2p_m & 2(1-p_m) \\ 0 & 0 & 1 \end{pmatrix} , \tag{A.4}$$

then, as in the case of Eqs. (A.1) and (A.2), the results may be combined into the following vector recursion

$$E\left[Z_n|Z_0 = \alpha\right] = E\left[Z_{n-1}|Z_0 = \alpha\right] M \ , \quad n \geq 1 \ , \tag{A.5}$$

where

$$E[Z_n|Z_0 = \alpha] \equiv (E[Z_{sn}|Z_0 = \alpha], E[Z_{mn}|Z_0 = \alpha], E[Z_{en}|Z_0 = \alpha]) \ .$$

Statistically fluent readers could obtain Eq. (A.5) via the route of conditional probability using the simple argument

$$E[Z_n|Z_0] = E\{E[Z_n|Z_{n-1}]\} \ ,$$

where we have used the Markov property to show that

$$E[Z_n|Z_0, Z_{n-1}] = E[Z_n|Z_{n-1}] = Z_{n-1}M \ .$$

Now, Eq. (A.5), if iterated, leads to the explicit representation of the desired mean as

$$E[Z_n|Z_0] = Z_0 M^n \ . \tag{A.6}$$

For p_s and/or $p_m = 0.5$, M^n can be readily obtained by direct multiplication. For general p_s and p_m, the spectral decomposition of M can be used to obtain M^n.

Rather than working with Eq. (A.6), we shall solve Eq. (A.5) via an approach suggested by J. Percus (personal communication) that exploits the recursive nature of the equation to develop an equivalent representation in terms of generating functions.

We define the generating functions

$$M_{\alpha\beta}(\lambda) = \sum_{n=0}^{\infty} E[Z_{\beta n}|Z_0 = \alpha]\lambda^n \ , \quad (\alpha, \beta = S, M, E) \ . \tag{A.7}$$

For the sake of illustration, suppose $\alpha = S$. Then after multiplying both sides of Eq. (A.5), with $\alpha = S$, by λ^n and summing over $n \geq 1$, we obtain for $\beta = S$

$$M_{ss}(\lambda) = 1 + 2p_s\lambda M_{ss}(\lambda) \tag{A.8}$$

since the $n = 0$ term $E[Z_{s0}|Z_0 = S] = 1$. Solving this equation we find

$$M_{ss}(\lambda) = \frac{1}{1 - 2p_s\lambda} = \sum_{n=0}^{\infty}(2p_s)^n\lambda^n \ , \tag{A.9a}$$

and thus by equating coefficients of λ^n in Eqs. (A.7) and (A.9a), we obtain

$$E[Z_{sn}|Z_0 = S] = (2p_s)^n \ . \tag{A.9b}$$

Similarly, we obtain

$$M_{sm}(\lambda) = 2(1 - p_s)\lambda M_{ss}(\lambda) + 2p_m\lambda M_{sm}(\lambda) \ , \tag{A.10a}$$

and

$$M_{se}(\lambda) = 2(1 - p_m)\lambda M_{sm}(\lambda) + \lambda M_{se}(\lambda) \tag{A.10b}$$

since $E[Z_{\beta 0}|Z_0 = S] = 0$ for $\beta = M, E$. Substituting Eq. (A.9a) into Eq. (A.10a) and solving we obtain

$$M_{sm}(\lambda) = \frac{2(1 - p_s)\lambda}{(1 - 2p_s\lambda)(1 - 2p_m\lambda)} \ . \tag{A.11a}$$

Further, substituting Eq. (A.11a) into Eq. (A.10b) and solving, we obtain

$$M_{se}(\lambda) = \frac{4(1 - p_m)(1 - p_s)\lambda^2}{(1 - 2p_s\lambda)(1 - 2p_m\lambda)(1 - \lambda)} \quad .$$

$(A.11b)$

The coefficient of λ^n in the expansion of Eqs. (A.11a) and (A.11b) gives $E[Z_{mn}|Z_0 = S]$ and $E[Z_{en}|Z_0 = S]$, respectively. For ease in expanding Eq. (A.11) in powers of λ, we apply the following rules as many times as necessary:

$$\frac{1}{(1 - a\lambda)(1 - b\lambda)} = \frac{1}{(a - b)} \left(\frac{a}{1 - a\lambda} - \frac{b}{1 - b\lambda} \right)$$

$$= \frac{1}{(a - b)} \sum_{n=0}^{\infty} (a^{n+1} - b^{n+1}) \lambda^n \quad , \quad a \neq b \quad , \quad (A.12a)$$

$$\frac{1}{(1 - \lambda)^2} = \sum_{n=0}^{\infty} (n + 1)\lambda^n \quad , \quad (A.12b)$$

and

$$\frac{1}{(1 - \lambda)^3} = \sum_{n=0}^{\infty} \frac{(n + 1)n}{2} \lambda^n \quad . \quad (A.12c)$$

The same technique leads to a set of equations for $M_{m\beta}(\lambda)$, $\beta = S, M, E$, in which $M_{ms}(\lambda)$ must be zero,

$$M_{mm}(\lambda) = 1 + 2p_m\lambda M_{mm}(\lambda) \quad , \quad (A.13a)$$

and

$$M_{me}(\lambda) = 2(1 - p_m)\lambda M_{mm}(\lambda) + \lambda M_{me}(\lambda) \quad . \quad (A.13b)$$

The case of $\alpha = E$ has $M_{es}(\lambda) = M_{em}(\lambda) = 0$ and

$$M_{ee}(\lambda) = 1 + \lambda M_{ee}(\lambda) \quad . \quad (A.13c)$$

Solving Eq. (A.13) and identifying the coefficients of λ^n by using Eq. (A.12) when necessary, we find the remaining expected values. We give the results below in the form of a summary.

(i) $E[Z_n|Z_0 = S] =$

$$
\begin{cases}
\left((2p_s)^n, \ \dfrac{(1-p_s)}{(p_s - p_m)}[(2p_s)^n - (2p_m)^n], \ \dfrac{2(1-p_s)(1-p_m)}{(p_s - p_m)} \times \right. \\[2mm]
\qquad \left. \left[\dfrac{(2p_s)^n}{(2p_s - 1)} - \dfrac{(2p_m)^n}{(2p_m - 1)} \right] + \dfrac{4(1-p_s)(1-p_m)}{(2p_s - 1)(2p_m - 1)} \right), \qquad p_s \neq p_m; p_s, p_m \neq 0.5 \\[4mm]
\left(1, \ \dfrac{1 - (2p_m)^n}{1 - 2p_m}, \ \dfrac{2(1-p_m)[n(1-2p_m) - 1 + (2p_m)^n]}{(1 - 2p_m)^2} \right), \qquad p_s = 0.5, p_m \neq 0.5 \\[4mm]
\left((2p_s)^n, \ \dfrac{2(1-p_s)[1 - (2p_s)^n]}{1 - 2p_s}, \right. \\[2mm]
\qquad \left. \dfrac{2(1-p_s)[n(1-2p_s) - 1 + (2p_s)^n]}{(1 - 2p_s)^2} \right), \qquad p_m = 0.5, p_s \neq 0.5 \\[4mm]
\left((2p)^n, \ 2n(1-p)(2p)^{n-1}, \right. \\[2mm]
\qquad \left. \dfrac{4(1-p)^2[1 - (2p)^n]}{(1 - 2p)^2} - \dfrac{4n(1-p)^2(2p)^{n-1}}{1 - 2p} \right), \qquad p_s = p_m = p \neq 0.5 \\[4mm]
\left(1, \ n, \ \dfrac{n(n-1)}{2} \right), \qquad p_s = p_m = 0.5 .
\end{cases}
$$

$$(A.14a)$$

(ii) $E[Z_n|Z_0 = M] =$

$$
\begin{cases}
\left(0, \ (2p_m)^n, \ \dfrac{2(1-p_m)[1 - (2p_m)^n]}{1 - 2p_m} \right), \qquad p_m \neq 0.5 \\[4mm]
(0, 1, n), \qquad p_m = 0.5.
\end{cases}
$$

$$(A.14b)$$

A.2   Variance

Now consider the variance of the process. In an irreducible branching process, explicit derivation of $var[Z_n|Z_0]$ is a complex task. However, the hierarchical structure of the $S \to M \to E$ process results in some simplification. The technique for obtaining the variance of the process is a natural extension of that for obtaining the mean. The definition of $var[Z_n|Z_0 = \alpha]$ is given by Eq. (2.15). Note that

$$var[Z_{\beta n}|Z_0] = E[Z_{\beta n}(Z_{\beta n} - 1)|Z_0] + E[Z_{\beta n}|Z_0] - (E[Z_{\beta n}|Z_0])^2 \qquad (A.15a)$$

and

$$cov[Z_{\beta n}, Z_{\beta' n}|Z_0] = E[Z_{\beta n} Z_{\beta' n}|Z_0] - E[Z_{\beta n}|Z_0]E[Z_{\beta' n}|Z_0] \ , \qquad (A.15b)$$

for $\beta, \beta' = S, M, E$.

We first develop recursions for $E[Z_{\beta n}(Z_{\beta n} - 1)|Z_0]$ and $E[Z_{\beta n} Z_{\beta' n}|Z_0]$ from Eqs. (2.8) and (2.9). For example, if $\alpha = S$, then

$$\left. \frac{\partial^2 f_s^{(n)}(\boldsymbol{\theta})}{\partial \theta_s^2} \right|_{\boldsymbol{\theta}=1} = E[Z_{sn}(Z_{sn} - 1)|Z_0 = S] \qquad (A.16)$$

by Eq. (2.8). Further, by Eq. (2.9),

$$\left. \frac{\partial^2 f_s^{(n)}(\boldsymbol{\theta})}{\partial \theta_s^2} \right|_{\boldsymbol{\theta}=1} = 2p_s^2 E[Z_{s,n-1}|Z_0 = S] + 4p_s^2 \ E[Z_{s,n-1}(Z_{s,n-1} - 1)|Z_0 = S]. \qquad (A.17)$$

Combining Eqs. (A.16) and (A.17), we obtain the recursion

$$E[Z_{sn}(Z_{sn-1})|Z_0 = S] = 2p_s^2 E[Z_{s,n-1}|Z_0 = S] + 4p_s^2 \ E[Z_{s,n-1}(Z_{s,n-1} - 1)|Z_0 = S] \ . \qquad (A.18)$$

We define generating functions

$$D_{\alpha\beta}(\lambda) = \sum_{n=0}^{\infty} E[Z_{\beta n}(Z_{\beta n} - 1)|Z_0 = \alpha]\lambda^n \ , \quad (\alpha, \beta = S, M, E) \ . \qquad (A.19)$$

Then multiplying both sides of Eq. (A.18) by λ^n and summing over $n \geq 1$ gives

$$D_{ss}(\lambda) = 2p_s^2 \lambda M_{ss}(\lambda) + 4p_s^2 \lambda D_{ss}(\lambda) \ . \qquad (A.20)$$

If we substitute for $M_{ss}(\lambda)$ from Eq. (A.9a) and solve for $D_{ss}(\lambda)$, we obtain

$$D_{ss}(\lambda) = \frac{2p_s^2 \lambda}{(1 - 2p_s\lambda)(1 - 4p_s^2\lambda)} \ . \qquad (A.21)$$

Applying Eq. (A.12) to obtain the coefficient of λ^n in the expansion of $D_{ss}(\lambda)$, we find

$$E[Z_{sn}(Z_{sn}-1)|Z_0 = S] = \begin{cases} \dfrac{p_s \cdot [(2p_s)^{2n} - (2p_s)^n]}{2p_s - 1} \ , & (p_s \neq 0.5) \\[2ex] \dfrac{n}{2} \ , & (p_s = 0.5) \ , \end{cases} \qquad (A.22)$$

from which an immediate consequence of Eqs. (A.14) and (A.15) is that

$$var[Z_{sn}|Z_0 = S] = \begin{cases} \dfrac{(1-p_s)\,[(2p_s)^{2n} - (2p_s)^n]}{2p_s - 1} \ , & (p_s \neq 0.5) \\[2ex] \dfrac{n}{2} \ , & \ \ \ \ \ \ \ (p_s = 0.5) \ . \end{cases} \qquad (A.23)$$

To continue the process and hence to obtain the full variance matrix of $Z_n|Z_0 = S$, we define generating functions for the product moments $E[Z_{\beta n}Z_{\beta' n}|Z_0 = S]$, $\beta \neq \beta'$, $\beta, \beta' = S, M, E$. Let

$$P_{\alpha\beta\beta'}(\lambda) = \sum_{n=0}^{\infty} E\,[Z_{\beta n}Z_{\beta' n}|Z_0 = \alpha]\,\lambda^n \ , \quad \beta \neq \beta' \ . \qquad (A.24)$$

Taking partial derivatives of Eqs. (2.8) and (2.9) once each with respect to θ_s and θ_m, and evaluating at $\theta = 1$, gives, for $\alpha = S$, the recursion

$$E[Z_{sn}Z_{mn}|Z_0 = S] = 2p_s(1 - p_s)\ E[Z_{s,n-1}|Z_0 = S]$$

$$+ 4p_s(1 - p_s)\ E[Z_{s,n-1}(Z_{s,n-1} - 1)|Z_0 = S] \qquad (A.25)$$

$$+ 4p_s p_m\ E[Z_{s,n-1}Z_{m,n-1}|Z_0 = S] \ .$$

After multiplying both sides of Eq. (A.25) by λ^n and summing over $n \geq 1$, we obtain

$$P_{ssm}(\lambda) = 2p_s(1 - p_s)\lambda\ M_{ss}(\lambda) + 4p_s(1 - p_s)\lambda\ D_{ss}(\lambda) + 4p_s p_m \lambda P_{ssm}(\lambda) \ .$$

Lastly, substituting for $M_{ss}(\lambda)$ from Eq. (A.9a) and $D_{ss}(\lambda)$ from Eq. (A.21), leads to

$$P_{ssm}(\lambda) = \frac{2p_s(1 - p_s)\lambda}{(1 - 2p_s\lambda)(1 - 4p_s p_m \lambda)} + \frac{8p_s^3(1 - p_s)\lambda^2}{(1 - 2p_s\lambda)(1 - 4p_s^2\lambda)(1 - 4p_s p_m \lambda)} \ . \qquad (A.26)$$

An expression for $P_{ssm}(\lambda)$ is necessary to be able to obtain $D_{sm}(\lambda)$, as the next (and last) example demonstrates. For $\alpha = S$, if we take second partial derivatives of Eqs. (2.8) and (2.9) with respect to θ_m, then set $\boldsymbol{\theta} = 1$, we obtain the recursion

$$E[Z_{mn}(Z_{mn} - 1)|Z_0 = S] = [2(1 - p_s)]^2 \, E[Z_{s,n-1}(Z_{s,n-1} - 1)|Z_0 = S]$$

$$+ 2(1 - p_s)^2 \, E[Z_{s,n-1}|Z_0 = S]$$

$$+ 2[2(1 - p_s)](2p_m) \, E[Z_{s,n-1}Z_{m,n-1}|Z_0 = S] \qquad (A.27)$$

$$+ (2p_m)^2 \, E[Z_{m,n-1}(Z_{m,n-1} - 1)|Z_0 = S]$$

$$+ 2p_m^2 \, E[Z_{m,n-1}|Z_0 = S] \ .$$

Hence the implicit expression for $D_{sm}(\lambda)$ becomes

$$D_{sm}(\lambda) = [2(1 - p_s)]^2 \lambda D_{ss}(\lambda) + (2p_m)^2 \lambda D_{sm}(\lambda)$$

$$+ 2(1 - p_s)^2 \lambda M_{ss}(\lambda) + 2p_m^2 \lambda M_{sm}(\lambda) \qquad (A.28)$$

$$2[2(1 - p_s)](2p_m)\lambda P_{ssm}(\lambda) \ ,$$

into which we substitute for $D_{ss}(\lambda)$, $M_{ss}(\lambda)$, $M_{sm}(\lambda)$, and $P_{ssm}(\lambda)$ from Eqs. (A.21), (A.9a), (A.11a), and (A.26) and then solve for $D_{sm}(\lambda)$. The coefficient of λ^n in the expansion of $D_{sm}(\lambda)$ gives $E[Z_{mn}(Z_{mn}-1)|Z_0 = S]$, which can then be used to determine $var[Z_{mn}|Z_0 = S]$ from Eq. (A.15). Note that the rule given by Eq. (A.12) can be used for ease in expansion of $D_{sm}(\lambda)$. In the above manner, one may obtain $var[Z_n|Z_0 = \alpha]$. We pursue this algebraically complex approach no further. Instead, complete sets of coupled equations defining the generating functions of $var[Z_n|Z_0 = \alpha]$ will be given below, following a statistical analysis to develop iterative expressions for the individual variances and covariances.

As in the case of $E[Z_n|Z_0 = \alpha]$, statistically fluent readers may prefer an alternative route to recursions similar to Eqs. (A.18), (A.25), and (A.27). We appeal to a standard result in statistical theory (cf. Mood, Graybill, and Boes, 1974) to show that

$$var[Z_{n+1}|Z_0] = E\{var \ [Z_{n+1}|Z_n]\} + var\{E[Z_{n+1}|Z_n]\} \ . \qquad (A.29)$$

To obtain $var[\mathbf{Z}_{n+1}|\mathbf{Z}_n]$, we define random variables $\mathbf{Y}^{(\alpha)} = (Y_s^{(\alpha)}, Y_m^{(\alpha)}, Y_e^{(\alpha)})$ to be the number of offspring of each type produced by a parent of type α, $\alpha = S, M, E$. The distributions of these random variables are:

$$\mathbf{Y}^{(s)} = \begin{cases} (2,0,0) & \text{with probability } p_s^2 \\ (1,1,0) & \text{with probability } 2p_s(1-p_s) \\ (0,2,0) & \text{with probability } (1-p_s)^2 \end{cases},$$

$$\mathbf{Y}^{(m)} = \begin{cases} (0,2,0) & \text{with probability } p_m^2 \\ (0,1,1) & \text{with probability } 2p_m(1-p_m) \\ (0,0,2) & \text{with probability } (1-p_m)^2 \end{cases},$$

$$\mathbf{Y}^{(e)} = (0,0,1) \quad \text{with probability } 1 .$$

Given $\mathbf{Z}_n$, the (random) population at time $(n+1)$ is a total of Z_{sn} random variables $\mathbf{Y}^{(s)}$, Z_{mn} random variables $\mathbf{Y}^{(m)}$, and Z_{en} random variables $\mathbf{Y}^{(e)}$. Because individuals are assumed to have stochastically independent fates and offspring distributions for all parents of a given type are identical,

$$var[\mathbf{Z}_{n+1}|\mathbf{Z}_n] = Z_{sn} \; var[\mathbf{Y}^{(s)}] + Z_{mn} \; var[\mathbf{Y}^{(m)}] + Z_{en} \; var[\mathbf{Y}^{(e)}] . \tag{A.30}$$

Now, the variance of $\mathbf{Y}^{(\alpha)}$ is a symmetric 3×3 matrix, namely

$$var[\mathbf{Y}^{(\alpha)}] = \begin{bmatrix} var[Y_s^{(\alpha)}] & cov[Y_s^{(\alpha)}, Y_m^{(\alpha)}] & cov[Y_s^{(\alpha)}, Y_e^{(\alpha)}] \\ & var[Y_m^{(\alpha)}] & cov[Y_m^{(\alpha)}, Y_e^{(\alpha)}] \\ * & & var[Y_e^{(\alpha)}] \end{bmatrix} .$$

We evaluate the elements of $var[\mathbf{Y}^{(\alpha)}]$ for each of the values of α. When $\alpha = S$, we obtain

$$var[\mathbf{Y}^{(s)}] = 2p_s(1-p_s) \begin{bmatrix} 1 & -1 & 0 \\ -1 & 1 & 0 \\ 0 & 0 & 0 \end{bmatrix} . \tag{A.31}$$

When $\alpha = M$, we obtain

$$var[\mathbf{Y}^{(m)}] = 2p_m(1-p_m) \begin{bmatrix} 0 & 0 & 0 \\ 0 & 1 & -1 \\ 0 & -1 & 1 \end{bmatrix} \quad . \tag{A.32}$$

Clearly, when $\alpha = E$,

$$var[\mathbf{Y}^{(e)}] = \begin{bmatrix} 0 & 0 & 0 \\ 0 & 0 & 0 \\ 0 & 0 & 0 \end{bmatrix} \quad , \tag{A.33}$$

since the offspring of a parent of type E is, with probability 1, a single type E cell.

We combine Eqs. (A.31)-(A.33) according to Eq. (A.30) to obtain

$$var[\mathbf{Z}_{n+1}|\mathbf{Z}_n] =$$

$$\begin{bmatrix} 2Z_{sn}p_s(1-p_s) & -2Z_{sn}p_s(1-p_s) & 0 \\ -2Z_{sn}p_s(1-p_s) & 2Z_{sn}p_s(1-p_s)+2Z_{mn}p_m(1-p_m) & -2Z_{mn}p_m(1-p_m) \\ 0 & -2Z_{mn}p_m(1-p_m) & 2Z_{mn}p_m(1-p_m) \end{bmatrix} \quad . \tag{A.34}$$

The third term in Eq. (A.29) may be rewritten using Eq. (A.5) as $var\{\mathbf{Z}_n\mathbf{M}\}$. Further, from the theory of linear models (cf. Searle, 1971), the following result holds:

$$var\{\mathbf{Z}_n\mathbf{M}\} = \mathbf{M}' \, var(\mathbf{Z}_n|\mathbf{Z}_0)\mathbf{M} \quad . \tag{A.35}$$

Substituting Eq. (A.35) into Eq. (A.29) gives

$$var[\mathbf{Z}_{n+1}|\mathbf{Z}_0] = E\left[var(\mathbf{Z}_{n+1}|\mathbf{Z}_n)\right] + \mathbf{M}' \, var(\mathbf{Z}_n|\mathbf{Z}_0)\mathbf{M} \quad , \quad n \geq 0 \quad . \tag{A.36}$$

If we introduce the notation,

$$var(\mathbf{Z}_n|\mathbf{Z}_0) = \begin{pmatrix} u_n & v_n & w_n \\ v_n & x_n & y_n \\ w_n & y_n & z_n \end{pmatrix} \quad , \tag{A.37}$$

Eq. (A.36) can be written as

$$
\begin{bmatrix} u_{n+1} \\ v_{n+1} \\ w_{n+1} \\ x_{n+1} \\ y_{n+1} \\ z_{n+1} \end{bmatrix} = \begin{bmatrix} \mu_{1n} \\ -\mu_{1n} \\ 0 \\ \mu_{1n} + \mu_{2n} \\ -\mu_{2n} \\ \mu_{2n} \end{bmatrix} +
$$

$$
\begin{bmatrix} 4p_s^2 & 0 & 0 & 0 & 0 & 0 \\ 4p_s(1-p_s) & 4p_s p_m & 0 & 0 & 0 & 0 \\ 0 & 4p_s(1-p_m) & 2p_s & 0 & 0 & 0 \\ 4(1-p_s)^2 & 8p_m(1-p_s) & 0 & 4p_m^2 & 0 & 0 \\ 0 & 4(1-p_s)(1-p_m) & 2(1-p_s) & 4p_m(1-p_m) & 2p_m & 0 \\ 0 & 0 & 0 & 4(1-p_m)^2 & 4(1-p_m) & 1 \end{bmatrix} \begin{bmatrix} u_n \\ v_n \\ w_n \\ x_n \\ y_n \\ z_n \end{bmatrix}
$$

$$(A.38)$$

where

$$\mu_{1n} = 2p_s(1-p_s)E[Z_{sn}|Z_0 = \alpha] \quad, \tag{A.39a}$$

$$\mu_{2n} = 2p_m(1-p_m)E[Z_{mn}|Z_0 = \alpha] \quad, \tag{A.39b}$$

and

$$u_0 = v_0 = w_0 = x_0 = y_0 = z_0 = 0 \quad . \tag{A.39c}$$

In the case $Z_0 = M$, $\mu_{1n} = 0, n \geq 0$, since a colony with a macrophage parent can never contain a stem cell. Therefore, as expected for this case, $u_n = v_n = w_n = 0$, $n \geq 0$. We can now continue as described above by multiplying both sides of Eq. (A.38) by λ^{n+1} and summing over $n \geq 0$, thus obtaining a system of linear equations in the generating functions for $var[Z_{\beta n}|Z_0]$ and $cov[Z_{\beta n}Z_{\beta' n}|Z_0]$, $(\beta, \beta' = S, M, E)$.

If we define

$$V_{\alpha\beta}(\lambda) = \sum_{n=0}^{\infty} var[Z_{\beta n}|Z_0 = \alpha]\lambda^n \quad, \quad (\beta = S, M, E) \tag{A.40a}$$

and

$$C_{\alpha\beta\beta'}(\lambda) = \sum_{n=0}^{\infty} cov[Z_{\beta n}, Z_{\beta' n}|Z_0 = \alpha]\lambda^n \quad , \quad (\beta, \beta' = S, M, E; \beta \neq \beta') \ , \qquad (A.40b)$$

then the complete sets of coupled equations for the generating functions of $var[Z_n|Z_0 = \alpha]$ expressed in matrix form are:

$\alpha = S$:

$$
\begin{bmatrix} V_{ss}(\lambda) \\ C_{ssm}(\lambda) \\ C_{sse}(\lambda) \\ V_{sm}(\lambda) \\ C_{sme}(\lambda) \\ V_{se}(\lambda) \end{bmatrix}
= 2p_s(1-p_s)\lambda M_{ss}(\lambda)
\begin{bmatrix} 1 \\ -1 \\ 0 \\ 1 \\ 0 \\ 0 \end{bmatrix}
+ 2p_m(1-p_m)\lambda M_{sm}(\lambda)
\begin{bmatrix} 0 \\ 0 \\ 0 \\ 1 \\ -1 \\ 1 \end{bmatrix}
$$

$$
+\lambda
\begin{bmatrix}
4p_s^2 & 0 & 0 & 0 & 0 & 0 \\
4p_s(1-p_s) & 4p_s p_m & 0 & 0 & 0 & 0 \\
0 & 4p_s(1-p_m) & 2p_s & 0 & 0 & 0 \\
4(1-p_s)^2 & 8p_m(1-p_s) & 0 & 4p_m^2 & 0 & 0 \\
0 & 4(1-p_s)(1-p_m) & 2(1-p_s) & 4p_m(1-p_m) & 2p_m & 0 \\
0 & 0 & 0 & 4(1-p_m)^2 & 4(1-p_m) & 1
\end{bmatrix}
\begin{bmatrix} V_{ss}(\lambda) \\ C_{ssm}(\lambda) \\ C_{sse}(\lambda) \\ V_{sm}(\lambda) \\ C_{sme}(\lambda) \\ V_{se}(\lambda) \end{bmatrix} ;
$$

$$\qquad\qquad (A.41a)$$

$\alpha = M$:

$$
\begin{bmatrix} V_{mm}(\lambda) \\ C_{mme}(\lambda) \\ V_{me}(\lambda) \end{bmatrix}
= 2p_m(1-p_m)\lambda M_{mm}(\lambda)
\begin{bmatrix} 1 \\ -1 \\ 1 \end{bmatrix}
$$

$$
+\lambda
\begin{bmatrix}
4p_m^2 & 0 & 0 \\
4p_m(1-p_m) & 2p_m & 0 \\
4(1-p_m)^2 & 4(1-p_m) & 1
\end{bmatrix}
\begin{bmatrix} V_{mm}(\lambda) \\ C_{mme}(\lambda) \\ V_{me}(\lambda) \end{bmatrix} \ ; \qquad\qquad (A.41b)
$$

$\alpha = E$:

$$V_{ee}(\lambda) = \lambda V_{ee}(\lambda) \ , \qquad\qquad (A.41c)$$

which has the trivial solution $V_{ee}(\lambda) = 0$.

The lower-triangular nature of the matrices in Eqs. (A.41a) and (A.41b) enables us to solve for a particular element of the vector of generating functions in terms of the elements higher in the vector. The hierarchical structure of the $S \to M \to E$ model thus leads to a straightforward (although rather tedious) solution for the variance of the Z_n process.

We shall not provide explicit, general solutions for $var[Z_n|Z_0]$; they are extremely complex. However, for particular values of p_s, p_m, and p_e, Eq. (A.41) can be solved numerically. This is the approach we took in Chapter 7. For the critical case ($p_s = 0.5 = p_m$), the algebra simplifies considerably and we obtain the following results:

(i) $var[Z_n|Z_0 = S]$ is composed of

$$var(Z_{sn}|Z_0 = S) = \frac{n}{2} \quad ;$$

$$cov(Z_{sn}, Z_{mn}|Z_0 = S) = \frac{n^2}{4} - \frac{3n}{4} \quad ;$$

$$cov(Z_{sn}, Z_{en}|Z_0 = S) = \frac{n^3}{12} - \frac{n^2}{2} + \frac{5n}{12} \quad ;$$

$$var(Z_{mn}|Z_0 = S) = \frac{n^3}{6} - \frac{n^2}{2} + \frac{5n}{6} \quad ;$$

$$cov(Z_{mn}, Z_{en}|Z_0 = S) = \frac{n^4}{16} - \frac{3n^3}{8} + \frac{7n^2}{16} - \frac{n}{8} \quad ;$$

$$var(Z_{en}|Z_0 = S) = \frac{n^5}{40} - \frac{5n^4}{24} + \frac{11n^3}{24} + \frac{5n^2}{24} - \frac{29n}{60} \quad . \qquad (A.42a)$$

(ii) $var[Z_n|Z_0 = M]$ is composed of

$$var(Z_{mn}|Z_0 = M) = \frac{n}{2} \quad ;$$

$$cov(Z_{mn}, Z_{en}|Z_0 = M) = \frac{n^2}{4} - \frac{3n}{4} \quad ; \qquad (A.42b)$$

$$var(Z_{en}|Z_0 = M) = \frac{n^3}{6} - \frac{3n^2}{4} + \frac{13n}{12} \quad .$$

Thus, as $n \to \infty$, the variance or covariance term will increase with the highest power of n in the appropriate polynomial.

By applying Eq. (3.3) to Eqs. (A.14) and (A.42), we can obtain explicit expressions in the strongly critical case for the mean and variance of the total colony size at the nth generation, $\sum_{\beta=S,M,E} Z_{\beta n}$. Thus

$$E\left[\sum Z_{\beta n}|Z_0 = S\right] = 1 + \frac{n}{2} + \frac{n^2}{2} \ ;$$

$$var\left[\sum Z_{\beta n}|Z_0 = S\right] = \frac{n^5}{40} - \frac{n^4}{12} + \frac{n^3}{24} + \frac{n^2}{12} - \frac{n}{15} \ ;$$

(A.43a)

and

$$E\left[\sum Z_{\beta n}|Z_0 = M\right] = 1 + n \ ;$$

$$var\left[\sum Z_{\beta n}|Z_0 = M\right] = \frac{n^3}{6} - \frac{n^2}{4} + \frac{n^2}{12} \ .$$

(A.43b)

The results of Eq. (A.43) were stated without proof in Macken et al. (1986).

Appendix B

Estimation of Branching Probabilities

We outline an approach that can lead to estimates of the branching probabilities, p_s and p_m. The approach depends upon using method of moment estimation techniques. In these techniques one equates sample moments (e.g., sample mean, sample variance) with their population equivalent (i.e., expected value, variance), which are theoretical values expressed in terms of the parameters of the model. If the model contains N unknown parameters, then N sample moments must be equated to their N population equivalents. The set of N equations is then solved (perhaps numerically) to produce estimates of the parameters.

In the present instance, data on colony growth is limited to measurements of total colony size, regardless of type of parent, at various times. Hence, method of moment estimation would require that we take moments of the total colony size and equate them to the corresponding moments of $Z_{sn} + Z_{mn} + Z_{en} = \sum_\alpha \eta_\alpha (Z_{sn} + Z_{mn} + Z_{en} | Z_0 = \alpha)$. In Appendix A we derive the first two moments of $(Z_n | Z_0)$, and consequently we can obtain the first two moments of $Z_{sn} + Z_{mn} + Z_{en}$. However, derivation of additional higher moments is too complex to be practical. Hence, from data on total colony size we can realistically estimate at most two parameters. We suggest that the logical choice of the parameters to estimate is p_s and p_m. Therefore, intelligent guesses must be made for the values of η_s, η_m, and η_e. Fortunately, as we indicated in Chapter 7, there are experimental estimates available that afford us some confidence in our "guesses."

To proceed with the estimation of p_s and p_m (assuming η_s, η_m, and η_e are known), we must compute the mean and variance of the total colony size. Since

$$E[Z_{sn} + Z_{mn} + Z_{en}] = E[Z_{sn}] + E[Z_{mn}] + E[Z_{en}] \ , \qquad (B.1)$$

we can use Eq. (2.19) for $E[Z_n]$ to obtain the expectation of the total colony size, where $E[Z_n | Z_0]$ is given by Eq. (2.14). Similarly,

$$var[Z_{sn} + Z_{mn} + Z_{en}] = var[Z_{sn}] + var[Z_{mn}] + var[Z_{en}]$$
$$+2 \ cov[Z_{sn}, Z_{mn}] + 2 \ cov[Z_{sn}, Z_{en}] + 2 \ cov[Z_{mn}, Z_{en}] \ . \tag{B.2}$$

Hence, we can use Eq. (2.20) for $var[\mathbf{Z}_n]$ along with the results of Appendix A for $var[\mathbf{Z}_n | \mathbf{Z}_0]$ to obtain the variance of the total colony size.

To complete the estimation process, suppose

$$\bar{Z}_n = \text{sample average colony size after } n \text{ generations}$$

and

$$s_{Z_n}^2 = \text{sample variance of the colony size after } n \text{ generations} \ .$$

We then find those values of p_s and p_m for which

$$\bar{Z}_n = E[Z_{sn} + Z_{mn} + E_{en}] \ , \tag{B.3}$$

and

$$s_{Z_n}^2 = var[Z_{sn} + Z_{mn} + Z_{en}] \ . \tag{B.4}$$

Clearly, the nonlinear nature of the expressions for the population moments will require a numerical solution of the set of Eqs. (B.3) and (B.4).

References

Absher, P. M., Absher, R. G.: Clonal variation and aging of diploid fibroblasts: Cinematographic studies of cell pedigrees. *Exp. Cell Res.* **103**, 247-255 (1976).

Alberts, B., Bray, D., Lewis, J., Raff, M., Roberts, K., Watson, J. D.: *Molecular Biology of the Cell.* New York: Garland, 1983.

Bell, E., Marek, L. F., Levinstone, D. S., Merrill, C., Sher, S., Young, I. T., Eden, M.: Loss of division potential in vitro: Aging or differentiation? *Science* **202**, 1158-1163 (1978).

Bjerknes, M.: Assessment of the symmetry of stem–cell mitoses. *Biophys. J.* **48**, 85-91 (1985).

Bjerknes, M.: A test of the stochastic theory of stem cell differentiation. *Biophys. J.* **49**, 1223-1227 (1986).

Burgess, A., Nicola, N.: *Growth Factors and Stem Cells.* Sydney: Academic Press, 1983.

Clark, S. C., Kamen, R.: The human hematopoietic colony-stimulating factors. *Science* **236**, 1229-1237 (1987).

Cowan, R., Morris, V. B.: Cell population dynamics during the differentiative phase of tissue development. *J. Theoret. Biol.* **122**, 205-224 (1986).

Day, R. S.: A branching-process model for heterogeneous cell populations. *Math. Biosci.* **78**, 73-90 (1986).

Dexter, T. M., Spooncer, E.: Growth and differentiation in the hemopoietic system. *Ann. Rev. Cell Biol.* **3**, 423-441 (1987).

Foster, J., Ney, P.: Limit laws for decomposable critical branching processes. *Z. Wahrscheinlichkeitstheorie verw. Gebiete* **46**, 13-43 (1978).

Fujimagari, T.: Controlled Galton-Watson process and its asymptotic behaviour. *Kodai Math. Sem. Rep.* **27**, 11-18 (1976).

Golub, E. S.: *Immunology: A Synthesis.* Sunderland, MA: Sinauer, 1987.

Good, P. I.: A stochastic model for in vitro aging. II. A theory of marginotomy. *J. Theoret. Biol.* **64**, 261-275 (1977).

Groopman, J. E.: Hematopoietic growth factors: from methylcellulose to man. *Cell* **50**, 5-6 (1987).

Gusella, J., Geller, R., Clarke, B., Weeks, V., Housman, D.: Commitment to erythroid differentiation by Friend erythroleukemia cells: A stochastic analysis. *Cell* **9**, 221-229 (1976).

Harris, T. E.: A mathematical model for multiplication by binary fission. In *The Kinetics of Cellular Proliferation*, F. Stohlman, Jr. ed. New York: Grune and Stratton, 1959.

Harris, T. E.: *The Theory of Branching Processes*. Berlin: Springer, 1963.

Hayflick, L.: The limited in vitro lifespan of human diploid cell strains. *Exp. Cell. Res.* **37**, 614-635 (1965).

Hayflick, L.: Cell biology of aging. *Fed. Proc.* **38**, 1847-1850 (1979).

Hayflick, L., Moorhead, P. S.: The serial cultivation of human diploid cell strains. *Exp. Cell Res.* **25**, 585-621 (1961).

Hodgson, G. S., Bradley, T. R.: Properties of haematopoietic stem cells surviving 5-fluorouracil treatment: Evidence for a pre-CFU-S cell? *Nature* **281**, 381-382 (1979).

Holliday, R., Huschtscha, L. I., Kirkwood, T. B. L.: Cellular aging: Further evidence for the commitment theory. *Science* **213**, 1505-1508 (1981).

Karlin, S., Taylor, H.M.: *A First Course in Stochastic Processes*, 2nd ed. New York: Academic Press, 1975.

Kesten, H., Stigum, B. P.: Limit theorems for decomposable multi-dimensional Galton-Watson processes. *J. Math. Anal. Applic.* **17**, 309-338 (1967).

Kirkwood, T. B. L., Holliday, R.: Commitment to senescence: A model for the finite and infinite growth of diploid and transformed human fibroblasts in culture. *J. Theoret. Biol.* **53**, 481-496 (1975).

Klebaner, F. C.: Population-size-dependent branching process with linear rate of growth. *J. Appl. Prob.* **20**, 242-250 (1983).

Klebaner, F. C.: Geometric rate of growth in population-size-dependent branching processes. *J. Appl. Prob.* **21**, 40-49 (1984).

Korn, A. P., Henkelman, R. M., Ottensmeyer, F. P., Till, J. E.: Investigations of a stochastic model of haemopoiesis. *Exp. Hemat.* **1**, 362-375 (1973).

Kurnit, D. M., Matthysse, S., Papayannopoulou, T., and Stamatoyannopoulos, G.: Stochastic branching model for hemopoietic progenitor cell differentiation. *J. Cell. Physiol.* **123**, 55-63 (1985).

Macken, C. A., Perelson, A. S., Stewart, C. C.: A stochastic model of macrophage colony growth. In: *Modelling of Biomedical Systems*, J. Eisenfeld and M. Witten, eds. New York: Elsevier, 1986, pp. 173-179.

Matsumura, T.: A rare family tree of cultured rat cells showing a change in proliferative potential. *Cell Biol. Int. Rep.* **7**, 931-935 (1983).

Matsumura, T.: Sequence of cell life phases in finitely proliferative population of cultured rat cells: A genealogical study. *J. Cell. Physiol.* **119**, 145-154 (1984).

Metcalf, D.: Clonal analysis of proliferation and differentiation of paired daughter cells: Action of granulocyte-macrophage colony-stimulating factor on granulocyte-macrophage precursors. *Proc. Natl. Acad. Sci. USA* **77**, 5327-5330 (1980).

Metcalf, D.: *The Hemopoietic Colony Stimulating Factors*. Amsterdam: Elsevier, 1984.

Mode, C. J.: *Multitype Branching Processes*. New York: Elsevier, 1971.

Mood, A. M., Graybill, F. A., Boes, D. C. : *Introduction to the Theory of Statistics*. New York: McGraw Hill, 1974.

Muller-Sieburg, C. E., Whitlock, C. A., Weissman, I. L. : Isolation of two early B lymphocyte progenitors from mouse marrow: A committed pre-pre-B cell and a clonogenic Thy-1^{lo} hematopoietic stem cell. *Cell* **44**, 653-662 (1986).

Nakahata, T., Gross, A. J., Ogawa, M.: A stochastic model of self-renewal and commitment to differentiation of the primitive hemopoietic stem cells in culture. *J. Cell. Physiol.* **113**, 455-458 (1982).

Nakahata, T., Ogawa, M.: Identification in culture of a class of hemopoietic colony-forming units with extensive capability to self-renew and generate multipotential colonies. *Proc. Natl. Acad. Sci. USA* **79**, 3843-3847 (1982).

Nedelman, J., Downs, H., Pharr, P.: Inference for an age-dependent, multitype branching-process model of mast cells. *J. Math. Biol.* **25**, 203-226 (1987).

Pharr, P. N., Nedelman, J., Downs, H. P., Ogawa, M., Gross, A. J.: A stochastic model for mast cell proliferation in culture. *J. Cell. Physiol.* **125**, 379-386 (1985).

Pharr, P. N., Suda, T., Bergmann, K. L., Avila, L. A., Ogawa, M.: Analysis of pure and mixed murine mast cell colonies. *J. Cell. Physiol.* **120**, 1-12 (1984).

Rubinow, S. I., Lebowitz, J. L.: A mathematical model of neutrophil production and control in normal man. *J. Math. Biol.* **1**, 187-226 (1975).

Searle, S. R.: *Linear Models*. New York: Wiley, 1971.

Souza, L. M., Boone, T. C., Gabrilove, J., Lai, P. H., Zsebo, K. M., Murdock, D. C., Chazin, V. R., Bruszewski, J., Lu, H., Chen, K. K., Barendt, J., Platzer, E., Moore, M. A. S., Mertelsmann, R., Welte, K.: Recombinant human granulocyte colony-stimulating factor: Effects on normal and leukemic myeloid cells. *Science* **232**, 61-65 (1986).

Spanier, J., Oldham, K. B.: *An Atlas of Functions*. New York: Hemisphere, 1987.

Steinberg, M. M., Brownstein, B. L.: Differentiation of cultured pre-adipose cells: A probability model. *J. Cell. Physiol.* Suppl. **2**, 37-50 (1982).

Stewart, C. C.: Formation of colonies by mononuclear phagocytes outside the bone marrow. In: *Mononuclear Phagocytes: Functional Aspects*, Part I, R. Van Furth, ed. The Hague and Boston: Martinus Nijhoff, 1980, pp. 377-413.

Stewart, C. C.: Regulation of mononuclear phagocyte proliferation. In: *The Reticuloendothelial System*, Vol. 7A, S. M. Reichard and J. P. Filkins, eds. New York: Plenum, 1984, pp. 37-56.

Suda, T., Suda, J., Ogawa, M.: Single-cell origin of mouse hemopoietic colonies expressing multiple lineages in variable combinations. *Proc. Natl. Acad. Sci. USA* 80, 6689-6693 (1983a).

Suda, T., Suda, J., Ogawa, M.: Proliferative kinetics and differentiation of murine blast cell colonies in culture: Evidence for variable G_0 periods and constant doubling rates of early pluripotent hemopoietic progenitors. *J. Cell. Physiol.* 117, 308-318 (1983b).

Suda, T., Suda, J., Ogawa, M.: Disparate differentiation in mouse hemopoietic colonies derived from paired progenitors. *Proc. Natl. Acad. Sci. USA* 81, 2520-2524 (1984).

Thomas, G. B., Jr.: *Limits*. Reading, Massachusetts: Addison-Wesley, 1963.

Till, J. E., McCulloch, E. A.: A direct measurement of the radiation sensitivity of normal mouse bone marrow cells. *Radiation Res.* 14, 213-222 (1961).

Till, J. E., McCulloch, E. A., Siminovitch, L.: A stochastic model of stem cell proliferation, based on the growth of spleen colony-forming cells. *Proc. Natl. Acad. Sci. USA* 51, 29-36 (1964).

Tukey, J. W.: A survey of sampling from contaminated distributions. In *Contributions to Probability and Statistics*, I. Olkin, S. Ghurye, W. Hoeffding, W. Madow and H. Mann, eds. Stanford, California: Stanford University Press, 1960, pp. 448-485.

Vogel, H., Niewisch, H., Matioli, G.: The self renewal probability of hemopoietic stem cells. *J Cell. Physiol.* 72, 221-228 (1968).

Vogel, H., Niewisch, H., Matioli, G.: Stochastic development of stem cells. *J. Theoret. Biol.* 22, 249-270 (1969).

von Melchner, H., Höffken, K.: Commitment to differentiation of human promyelocytic leukemia cells (HL60): An all-or-none event preceded by reversible losses of self-renewal potential. *J. Cell. Physiol.* 125, 573-581 (1985).

Wichmann, H.-E., Loeffler, M.: *Mathematical Modeling of Cell Proliferation: Stem Cell Regulation in Hemopoiesis*. Vols. 1 and 2. Boca Raton, Florida: CRC Press, 1985.

Your source for advances in theoretical biology and biomathematics

ISSN 0303-6812   Title No. 285

For mathematicians and biologists working in a wide variety of fields – genetics, demography, ecology, neurobiology, epidemiology, morphogenesis, cell biology – **the Journal of Mathematical Biology** publishes:

- papers in which mathematics is used for a better understanding of biological phenomena
- mathematical papers inspired by biological research, and
- papers which yield new experimental data bearing on mathematical models

Abstracted/Indexed in: Current Contents, Excerpta Medica, Index Medicus, Mathematical Reviews, Science Abstracts, Animal Breeding Abstracts, Compumath, Helminthological Abstracts, Index to Scientific Reviews, Plant Breeding Abstracts, Zentralblatt für Mathematik

Subscription Information:
To enter your subscription, or to request sample copies, contact Springer-Verlag, Dept. ZSW, Heidelberger Platz 3, D-1000 Berlin 33, W. Germany

Springer-Verlag
Berlin Heidelberg New York
London Paris Tokyo Hong Kong

Springer

Bio-mathematics

Managing Editor: S. A. Levin

Editorial Board: M. Arbib,
H. J. Bremermann, J. Cowan,
W. M. Hirsch, J. Karlin,
J. Keller, K. Krickeberg,
R. C. Lewontin, R. M. May,
J. D. Murray, A. Perelson,
T. Poggio, L. A. Segel

Volume 15
D. L. DeAngelis, W. M. Post, C. C. Travis

Positive Feedback in Natural Systems

1986. 90 figures. XII, 290 pages. ISBN 3-540-15942-8

Contents: Introduction. – The Mathematics of Positive Feedback. – Physical Systems. – Evolutionary Processes. – Organisms Physiology and Behavior. – Resource Utilization by Organisms. – Social Behavior. – Mutualistic and Competitive Systems. – Age-Structured Populations. – Spatially Heterogeneous Systems: Islands and Patchy Regions. – Spatially Heterogeneous Ecosystems; Pattern Formation. – Disease and Pest Outbreaks. – The Ecosystem and Succession. – Appendices. – References. – Subject Index. – Author Index.

Volume 16

Complexity, Language, and Life: Mathematical Approaches

Editors: J. L. Casti, A. Karlqvist
1986. XIII, 281 pages. ISBN 3-540-16180-5

Contents: Allowing, forbidding, but not requiring: a mathematic for human world, – A theory of stars in complex systems. – Pictures as complex systems. – A survey of replicator equations. – Darwinian evolution in ecosystems: a survey of some ideas and difficulties together with some possible solutions. – On system complexity: identification, measurement, and management. – On information and complexity. – Organs and tools; a common theory of morphogenesis. – The language of life. – Universal principles of measurement and language functions in evolving systems.

Volume 17

Mathematical Ecology

An Introduction
Editors: Th. G. Hallam, S. A. Levin
1986. 84 figures. XII, 457 pages. ISBN 3-540-13631-2

Contents: Introduction. – Physiological and Behavioral Ecology. – Population Ecology. – Communities and Ecosystems. – Applied Mathematical Ecology. – Author Index. – Subject Index.

Volume 18

Applied Mathematical Ecology

Editors: S. A. Levin, T. G. Hallam, L. J. Gross
1988. ISBN 3-540-19465-7. In preparation

Volume 19
J. D. Murray

Mathematical Biology

1988. ISBN 3-540-19460-6. In preparation

Springer-Verlag
Berlin Heidelberg New York
London Paris Tokyo Hong Kong

Springer